RANDOMISED CONTROLLED CLINICAL TRIALS, Second Edition

RANDOMISED CONTROLLED CLINICAL TRIALS, Second Edition

Christopher J Bulpitt
Division of Geriatric Medicine
Royal Postgraduate Medical School
Hammersmith Hospital
Du Cane Road
London W12

KLUWER ACADEMIC PUBLISHERS
Boston / Dordrecht / London

Distributors for North America:
Kluwer Academic Publishers
101 Philip Drive
Assinippi Park
Norwell, Massachusetts 02061 USA

Distributors for all other countries:
Kluwer Academic Publishers Group
Distribution Centre
Post Office Box 322
3300 AH Dordrecht, THE NETHERLANDS

Library of Congress Cataloging-in-Publication Data

A C.I.P. Catalogue record for this book is available
from the Library of Congress.

Printed on acid-free paper.

Printed in the United States of America

CONTENTS

PREFACE

Bradford Hill has defined a clinical trial as *"A carefully and ethically designed experiment with the aim of answering some precisely framed question"* [1]. This definition specifies a careful design and requires the provision of adequate controls. Random allocation of treatments to subjects is important to ensure that the treated and control groups are similar. Therefore this book is entitled *Randomised Controlled Clinical Trials*. We can define a randomised controlled trial by rewriting Bradford Hill's definition as follows, *"A carefully and ethically designed experiment which includes the provision of adequate and appropriate controls by a process of randomisation, so that precisely framed questions can be answered."*

I am a firm advocate of Randomised Controlled Clinical Trials but intend to give a balanced view of the advantages and disadvantages of these ethical experiments. This book is directed primarily at the medical research worker, although certain chapters may find a wider application.

When discussing a randomised controlled trial, it is neither practicable nor desirable to divorce theory from practice, however the first seven chapters concentrate mainly on theory, and the remainder focus on practice. The segment on trial design is followed by sections on writing the protocol, designing the forms, conducting the trial, and analysing the results. This book is meant to serve both as a reference manual and a practical guide to the design and performance of a trial.

Preface to the second edition

After 13 years there are new areas to discuss and more recent trials to be included. Good Clinical Practice; evaluation of quality of life; measurement of the benefit : risk comparison; determination of cost-effectiveness and cost utility; stopping rules for trials; meta analysis and subgroup analysis are all new sections. The references are expanded from 305 to 512 and include the recent advances in trial design, such as the n-of-1 trials and megatrials, and up-to-date examples to illustrate the points made in the 20 chapters.

ACKNOWLEDGEMENTS

I am deeply indebted to many persons who have provided invaluable assistance in the preparation of this second edition. Miss Alison Palmer kindly read and corrected the manuscript. My secretary, Miss Nicola Piggott, patiently processed and printed the several versions of the work. Nevertheless I remain responsible for any errors.

1. INTRODUCTION

DEFINITION

A randomised controlled trial was defined in the preface as *"A carefully and ethically designed experiment which includes the provision of adequate and appropriate controls by a process of randomisation, so that precisely framed questions can be answered."* In medical research, treatment is allocated to subjects or certain periods of time by a random (chance) procedure.

WHY PERFORM A RANDOMISED CONTROLLED TRIAL?

The primary objective in writing this book was to demonstrate the importance of performing randomised controlled trials and the second was to help in the design and performance of such trials. We do not expect to convince the reader of the necessity for randomised controlled trials in a brief introduction as an entire half of the book is intended for this; but a preliminary discussion may be appropriate. Any reader already convinced of the necessity for randomised clinical trials should proceed to the next section.

In the current medical literature, opinion is often held in high esteem and randomised controlled trials constitute only a small proportion of research reports. Articles consisting of observations without randomised comparison groups can be valuable and often generate hypotheses, some of which are subsequently tested by randomised controlled trials, but the preponderance of observational studies over controlled experimentation is surprising. We leave the reader to imagine the number of controlled trials to be found in literature on the social *sciences* [our italics].

Cochrane [2] observed that not only were randomised controlled trials neglected in other fields but that in medicine these trials are carried out in developed, capitalist, predominantly Protestant countries. He understood why underdevelopment mitigated against such trials and could only speculate why Communist or Roman Catholic countries should inhibit the performance of randomised controlled trials. This association with religous belief is rapidly disappearing, as any review of the increasing Italian literature will reveal.

Every time a treatment is prescribed for a patient, whether pharmaceutical agent, operation, diet, psychological counselling, physiotherapy, or other health care strategy, the medical practitioner is conducting a trial of treatment in that patient. Similarly, when an administrator organises health care for a community, for example screening, immunisation, or better housing, an experiment or trial is performed. However, we wish to know whether or not the experiment works and it is not sufficient to observe that the health of the patient or community improves as such an improvement may have nothing to do with the experiment. Patients may get better without treatment and the health of a community may improve without screening for diseases. We therefore need controls who do not receive the intervention. These controls provide the baseline against which treatment or intervention can be assessed: hence the term *controlled trials*. The control groups should be the same as the intervention group in all respects apart from the intervention procedure. Chapter 7 is devoted to demonstrating that controls must be observed concurrently with the intervention group and that those treated and those serving as controls must be determined by chance alone. The allocation by chance is known as randomisation, leading to the term *randomised controlled trial* or RCT. The allocation can be performed by tossing a coin on each occasion or more usually by the use of random number tables (chapter 7). Tossing a coin is open to manipulation and is not advised.

Trials of new therapy

Randomised controlled trials are necessary to

prove the effectiveness of new health care strategies or treatment and to prevent the introduction of new but useless treatments.

Trials of accepted treatment

Many currently accepted treatments require proof of their effectiveness and trials are still necessary. Such treatments fall into three groups: first, a therapy may have been introduced prior to the advent of clinical trials. Cochrane [2] suggests that psychotherapy, physiotherapy, and surgery for carcinoma of the bronchus can be included in this group. Second, there may be experimental proof of biochemical, psychological, or other effect, but no evidence that the treatment does more good than harm as long-term therapy. Examples are provided by aspirin for the primary prevention of myocardial infarction and oral hypoglycaemic drugs for maturity-onset diabetes mellitus (chapter 19). Third, a treatment may have been subjected to randomised controlled trials, but the results are equivocal. Cochrane suggests tonsillectomy as an example of this group.

The place of randomised controlled trials

The place of randomised controlled trials must not be exaggerated. The advantages and disadvantages of clinical trials are discussed further in chapter 20. There was no necessity for a trial of streptomycin in tuberculous meningitis; one survival in an otherwise uniformly fatal condition was very conclusive! However, randomised controlled trials of streptomycin were useful in pulmonary tuberculosis [3]. These trials could have delayed the introduction of active treatment, but a shortage of streptomycin proved to be the limiting factor. A delay may also be of value when the treatment proves to be of no benefit. Trials may produce erroneous results but these occur far less frequently than in uncontrolled observational studies and, the quality of medical care given to patients in clinical trials is much higher than for the usual processes of medical care [2].

DESCRIPTION OF THIS BOOK

This book is intended as a reference manual for research workers involved in randomised controlled trials and is aimed at the field of medical research. It is hoped that the contents of the book are also relevant to the needs of dental, veterinary, and social science research workers. Many aspects of trial design have also been employed in agricultural experiments.

After this introduction, the book considers historical aspects of the subject and the ethical aspects of trial design. We have first to agree the ground rules for trials in patients and in normal volunteers. What risks are allowable for the former, if any? It has been suggested that it would be unethical to perform a trial involving the transmission of infectious hepatitis in man, yet such a trial was carried out.

Following the discussion on ethics, subsequent chapters will closely define the trial objectives, the different trial designs, the number of subjects required, randomisation, and freeing observations from bias. The next section of the book is concerned with very practical matters: writing the protocol, recruitment of subjects, information to be collected, conduct of the trial, stopping rules, analysing the results, and variability and validity of the results. These considerations are followed by the evaluation of subjective well-being, measurement of quality of life, the detection of adverse reactions, measurement of the benefit : risk ratio, cost-effectiveness and cost utility, trials on new drugs, reporting results, why results are not always accepted, and the advantages and disadvantages of randomised controlled trials.

It is hoped that readers who are not involved in the performance of RCTs will find themselves better able to assess the results of such trials. More importantly, promotional studies, masquerading as important randomised trials, may thus be given the scant attention they deserve. We will attempt to identify the necessary qualities of a satisfactory trial so that we may more readily assess the results. Initially we will ask if the trial incorporates satisfactory controls. A trial that is not controlled is often called a study and the results most favourable to a particular treatment usually derive from an open evaluation. An example is given by an article sent to general practitioners by a pharmaceutical company making drug X and entitled *"Drug X in hypertension. General Practice Study. Preliminary report on 717 patients treated originally with methyldopa."* The dose of methyldopa in these patients was reduced but not stopped and drug X was started. The article stated *"Whatever the reasons, there is good evidence that in the majority of this group, control of the blood pressure improved when the dose of methyldopa was reduced and drug X was substituted. After drug X was introduced there was a reduction in unwanted side effects, and four out of five patients reported subjective improvement."* This was an open evaluation with a predictable result. Give a new drug with enthusiasm and the patient will feel better. This study made nearly 600 patients feel better, but some or all of the improvement may have been due to the attention of the general practitioner rather than the new drug.

An article in the *Sunday Times* of January 29, 1978 attacked these marketing trials and put the situation clearly. *"If the doctor believes the new drug may help, he will almost certainly tell the patient, so the patient will be inclined to prefer it to this previous drug, which may be just as good. Few of these trials make a proper scientific comparison of the new drug with other drugs or with a dummy tablet."* The newspaper article went on to consider the profits to be made by pharmaceutical companies who introduce new drugs to general practitioners in this manner as the drug continues to be prescribed after the end of the trial. Post marketing surveillance is also employed to advertise and promote new drugs, the uncontrolled nature of such studies limits the value of the data provides on adverse drug reactions. Not every one agrees that making a profit is undesirable, especially in view of the therapeutic advances made by the pharmaceutical industry. However, there must be adequate proof of benefit and the randomised controlled trial is the method of choice.

2.   THE HISTORY OF CONTROLLED TRIALS

THE EARLIEST TRIALS

Archie Cochrane popularised the term Randomised Controlled Trial [2] but RCTs date back to the 1940s [3,4] when the Medical Research Council reported trials on the use of streptomycin in pulmonary tuberculosis [5] and the use of whooping-cough vaccination [6]. The first well documented randomised controlled trial of medical treatment may have been that organised by the Medical Research Council and reported in 1948 [3]. However, Armitage has described a possible RCT with group randomisation dating from 1662 [3] and R A Fisher introduced randomised trials in agricultural research in 1926 [3].

Non-randomised trials date back many years. L'Etang [7] considered that the story of Daniel contained a report of clinical trial. Nebuchadnezzar II organised a trial by giving youths of royal blood, including Daniel, a rigid diet of meat and wine for three years. The trial was supervised by a eunuch [monitor]. Nebuchadnezzar's trial was not controlled but Daniel *"persuaded the monitor to give him and three others a diet of pulse and water for 10 days."* L'Etang reported that these four were *"fairer in countenance and fatter in body than the other subjects who were given meat and wine"* and concluded that *"Daniel had ruined the trial ... and the trial had become uncontrolled."* Daniel had not ruined the trial but had performed one of the first *controlled* trials: a within-subject cross-over study.

Bull has reviewed the history of clinical trials and we are indebted to him for much material in this chapter [8]. Bull cited a second unintentional trial by Ambrose Paré. In 1537 Paré was responsible for the treatment of numerous wounded and ran out of boiling oil used for cauterising the wounds. He was *"constrained to apply in its place a digestive made of yolks of eggs, oil of roses and turpentine."* The following day he was surprised to find that those receiving the new medicant *"feeling but little pain, their wounds neither swollen nor inflamed..."* Those who received boiling oil *"were feverish with much pain and swelling about their wounds."* Paré concluded that the digestive was superior to burning oil but perhaps we would now suggest a longer period of observation would be appropriate in view of the likelihood of subsequent sepsis.

Scurvy

Bull also reported an unintentional trial from 1600 on an expedition to India by the East India Company. Only one of four ships had lemon juice provided. The ship in question was almost free from scurvy yet the condition was rampant on the other three ships. The company provided lemon juice on all its ships thereafter but presumably this preventive treatment was not fully accepted until 150 years later, when James Lind performed a controlled trial. Bradford Hill quotes James Lind in his book *Statistical Methods in Clinical and Preventive Medicine* [9].

On the 20th May, 1747, I took twelve patients in the scurvy, on board the Salisbury at sea. Their cases were as similar as I could have them. They all in general had putrid gums, the spots and lassitude, with weakness of their knees. They lay together in one place, being a proper apartment for the sick in the fore-hold; and had one diet common to all, viz water-gruel sweetened with sugar in the morning, fresh mutton-broth often times for dinner; at other times puddings, boiled biscuit with sugar etc and for supper, barley and raisins, rice and currants, sago and wine, or the like. Two of these were ordered each a quart of cyder a day. Two others took twenty-five gutts of elixir vitriol three times a day, upon an empty stomach; using a gargle strongly acidulated with it for their mouths. Two others took spoonfuls of vinegar three times a day, upon an empty stomach; having their gruels and their other food well acidulated with it, as also the gargle for their mouths. Two of the worst patients, with the tendons in the ham rigid (a symptom none of the rest had) were put under a course of sea-water. Of this they drank half a pint every day, and sometimes more or less as it operated by way of gentle physic. Two others had each two oranges and one lemon given them every day. These they ate with greediness, at different times, upon an empty stomach. They continued but six days under this course, having consumed the quantity that could be spared. The two remaining patients, took the bigness of a nutmeg three times a day of an electuary recommended by a hospital-surgeon, made of garlic, mustard-feed, rad raphan, balsam of Peru, and gum myrrh; using for common drink barley-water well acidulated with tamarinds; by a decoction of which, with the addition of cremor tartor, they were greatly purged three or four times during the course.

The consequence was, that the most sudden and visible good effects were perceived from the use of the oranges and lemons, one of those who had taken them, being at the end of six days fit for duty. The spots were not indeed at that time quite off his body, nor his gums sound, but without any other medicine, than a gargle of elixir vitriol, he became quite healthy before we came into Plymouth, which was on the 16th June. The other was the best recovered of any in his condition; and being now deemed pretty well, was appointed nurse to the rest of the sick.

James Lind showed the superiority of citrus fruits in the treatment of scurvy. Interestingly, he commits one fundamental error: he appears to have given two of the worst patients a particular treatment (sea water). Perhaps sea water was his favourite treatment. If this was the case, and we assume that more severely affected patients are less easy to cure, the provision of a favourite treatment for these patients will mitigate against demonstrating a benefit. Random allocation to treatment groups prevents this difficulty; this advance did not occur until much later.

VACCINATION AGAINST SMALLPOX

In the early eighteenth century inoculation against smallpox with live virus was introduced from Constantinople by Maitland and Lady Mary Wortley Montagu. (They, of course, knew nothing about viruses and reported the inoculation of "smallpox matter".) They arranged for six convicts to be inoculated; all survived and later one was exposed to smallpox and found to be immune [10]. This trial did not reveal that the use of live virus could frequently lead to death and that those inoculated could be infectious and lead to an increase of smallpox in the community.

A similar trial, but using cowpox matter, was next performed by a farmer, Benjamin Jesty in 1774 [11]. Country folk reported that if they had cowpox they would not get smallpox and that cowpox was a mild disease. Farmer Jesty vaccinated himself, his wife, and two children with cowpox material using a stocking needle. Apparently the children were later inoculated with smallpox and were unaffected. Mr Jesty proceeded to vaccinate his milkmaids but his neighbours considered such a "bestial" manifestation of smallpox should not be given to man! Benjamin Jesty countered that if we were prepared to eat beef and drink milk, we could be vaccinated with cowpox.

Jesty had performed a remarkable trial and 20 years later Edward Jenner, not knowing of farmer Jesty, was considering the problem and remembered the words of his teacher, the famous surgeon, John Hunter: "Don't think. Do an experiment." Jenner proceeded to do a similar trial which he wrote up in 1798 in *An Inquiry into the Causes and Effects of the Variolate Vaccine* [12]. He described 23 patients with cowpox who were resistant to smallpox inoculation and assumed that all persons who had neither contracted cowpox nor smallpox would react positively to inoculation with smallpox matter. The trial was not controlled in the strict sense of the word, and controls *were necessary* owing to the possibility of either natural immunity or previously acquired immunity. Such persons could be vaccinated with cowpox and be immune to smallpox without cause and effect.

Jenner was partly aware of this problem and stated, *"To convince myself that the variolus matter made use of was in a perfect state, I at the same time inoculated a patient with some of it who had never gone through the cowpox, and it produced the smallpox in the usual manner."* A further problem arose from interpreting resistance to the disease. We now know that inoculating with smallpox would be unethical as a control procedure. Pearson [13] was less interested in the resistance of his subjects than the state of his variolus matter and made further controlled observations, also in 1798. He observed three patients who had had cowpox and two who had not. Only the three who had had cowpox were immune to inoculation. Interestingly and commendably, Pearson recommended *"well-directed observation in a thousand cases of inoculated cowpox."* Waterhouse performed a similar controlled trial and published two years later [14]. He employed the same number of controls (two) but had 12 in his treatment group.

THE USE OF PLACEBOS IN THE NINETEENTH CENTURY

When there is doubt about the effectiveness of a particular treatment or when effective treatment is either not available or not required in the short-term, the modern controlled trial may employ a period of placebo medication. In 1801 Haygarth was one of the first to employ placebo treatment [15]. He used dummy appliances to investigate the effects of Perkin's tractors. Not only was the trial possibly the first to use placebo medication, but Haygarth quoted Lind, thus *"an important lesson in physic is here to be learnt, viz the wonderful and powerful influence of the passions of the mind upon the state and disorders of the body. This is too often overlooked in the cure of disease ..."*
In 1865 Sutton published a trial of mint water by Dr Gull in 20 patients with rheumatic fever [16]. He used mint water, not as an active but as a placebo treatment. On observing a marked tendency to spontaneous cure he remarked, *"the best treatment for rheumatic fever has still to be determined."* He also reported, *"No selection was made, but that Dr Gull treated the cases which happened to be admitted into his wards on the same plan; and we would further beg to say that these reports were not kept for any special object, nor are they as complete as they might be; yet the facts stated, may be fully relied upon, and so far answer our purpose."*

These interesting admissions make us suspect that the study was not so well conceived and prospective in design as it first appeared. However, the honesty of Dr Sutton led to further qualifications that we now agree would be unlikely to influence the course of the disease. *"... these cases cannot be considered to have been treated solely on the expectant plan, for an occasional dose of Dover's powder or half a grain of opium, right or wrong, and two or three ounces of brandy a day, are remedies that might be fairly expected to exercise some, although, perhaps, little influence over the course of the disease."* He wisely stated, *"Therefore, cases treated, as the following cases have been, by such simple means that we might almost consider them to be unassisted by any remedy, are invested with no little interest ... the results ... will probably warrant us concluding that we ought not to be too hasty in considering the apparent sudden and favourable change in the symptoms due to any medicine administered."*

SCIENTIFIC BASIS FOR CONTROLLED TRIALS

Although the advent of the randomised controlled trial had to wait until the twentieth century, much of the related scientific thinking was published in the nineteenth century. Laplace thought that probability theory ought to be extended to help explain the results observed in medical practice [17], and P C A Louis advocated numeracy in assessing results [18]. Louis stated, *"As to different methods of treatment, if it is possible for us to assure ourselves of superiority of one another among them ... it is doubtless to be done by enquiring if ... a greater number of individuals have been cured by one means than another. Here again it is necessary to count."* He went on to consider the necessity for controls *"in order that the calculation may lead to useful or true results ... we ought to know the natural progress of the disease."* He also appeared concerned about noncompliance: *"we ought to know ... whether the subjects have not committed errors of regimen."* Amusingly, he thought his numerical method offered *"real difficulties in its execution ... this method requires much more labour and time than the most distinguished members of our profession can dedicate to it. But what signifies this reproach, except that the research of truth requires much labour, and is beset with difficulty."*

Louis used his numerical method in investigating the effect of venesection in 78 cases of pneumonia [19]. Some patients were bled and others were not. Louis not only examined mortality but also symptoms and signs and concluded that bleeding made no difference in outcome. This result was not in keeping with the medical practice at this time, and not unexpectedly, caused an uproar. However, his findings came to be accepted, a triumph for the clinical trial.

THE PROVISION OF HISTORICAL CONTROL GROUPS

By the middle of the nineteenth century, rigorous methods of observation had been defined, the necessity for controls realised, and even the statistical theory of probability could have been used in the analysis of results. However, the selection of controls still led to biased results. Elisha Bartlett [20] described the essential requirements for control and treated patients: they should

1. have equal disturbing factors of location, social class, and the like
2. be susceptible of a clear and positive diagnosis
3. not be selected
4. be subjected to a clearly defined method of treatment.

The provision of a more appropriate control group came to be recognised as important and trials started to employ carefully followed historical controls. In 1870 Lister [21] compared the mortality of 35 historical controls with 40 patients treated with antiseptics. Forty-three percent of the controls died but only 15 percent of the treated group. Bull [8] first pointed out that Lister was cautious about drawing conclusions and, second, that more appropriate controls *"might have prevented the bitter and profitless controversy which raged for many years."* Lister stated, *"These numbers are, no doubt, too small for a satisfactory statistical comparison ..."* Bull pointed out that, *"The chi-squared test shows them to be highly significant"*; perhaps the controversy would have raged less if Lister were a more effective lecturer and a more dogmatic writer. However, historical controls are not appropriate for randomised controlled trials. Pocock [22] has listed 19 instances of the same intervention being used twice in consecutive groups of patients in the same institution. In four of these instances mortality was significantly different between the groups.

THE PROVISION OF CONCURRENT CONTROL GROUPS

The lack of acceptance for Lister's trial can be contrasted with that of Pasteur's vaccine for the prophylaxis of anthrax in animals. Pasteur used 60 sheep in the experiment; 25 were inoculated and then infected and 25 were not inoculated but were infected. An additional ten sheep were neither inoculated nor infected. Chance allocation appears to have been employed to some extent in this trial as critical observers suggested the order in which pairs of inoculated and control animals should be infected [23]. All the animals who had been inoculated survived; the 25 controls died. The results of this trial were immediately accepted.

Another early controlled trial was performed by Fibiger. In 1898 he reported a trial of anti-diphtheria serum in alternate patients [24]. He studied 488 patients and showed a reduction in mortality in the patients treated with serum. He also recorded the fact that the diphtheritic membrane disappeared quicker in the treated cases. Armitage [3] considered this trial to be the first example of allocation by *alternation* although Karl Pearson advocated its use for a trial of typhoid immunization in 1904 [25] and the Medical Research Council employed this design in the 1930s [26].

In 1945 a trial of penicillin in the treatment of wounds was attempted in the 21 Army Group [27]. The control group was to be those who were given any alternative treatment. Unfortunately, the surgeons were unwilling to withhold penicillin in the presence of serious wounds and the group treated with penicillin were more seriously affected. Despite this bias, the wounds healed quicker in the penicillin-treated group. The potential is large for biased allocation when the next treatment to be given is known. Random allocation is therefore required.

THE DELIBERATE USE OF RANDOMISATION TO PRODUCE SIMILAR TREATMENT AND CONTROL GROUPS

Peirce and Jastrow advocated randomisation in 1884 for experiments in psychology [28], R A Fisher from 1926 in agricultural experiments [29] and Amberson and colleagues employed group randomisation for gold therapy for patients with tuberculosis in 1931 [30].

One of the first modern trials to deliberately employ individual randomisation was the Medical Research Council trial of streptomycin reported in 1948 [5]. the introduction to the trial pointed out that the natural history of pulmonary tuberculosis was so variable that *"evidence of improvement or cure following the use of a new drug in a few cases cannot be accepted as proof of the effect of that drug."* The introduction further pointed out that there had been only one report of an adequately controlled trial in tuberculosis and that this trial was negative and counteracted the exaggerated claims for gold treatment that had been made for over 15 years [30]. Patients entering the trial of streptomycin were restricted to those who were both unlikely to improve spontaneously and yet were likely to respond to an active chemotherapeutic agent. It was therefore decided that patients chosen had to have acute progressive bilateral tuberculosis; subjects were excluded if they had long-standing disease. The control treatment was to be bed rest and patients were excluded if they required pulmonary-collapse therapy.

The new feature of the trial was the individual randomisation of patients into control and treated groups. The report stated:

Determination of whether a patient would be treated by streptomycin and bed-rest (S case) or by bed-rest alone (C Case) was made by reference to a statistical series based on random sampling numbers drawn up for each sex at each centre by Professor Bradford-Hill; the details of the series were unknown to any of the investigators or to the coordinator and were contained in a set of sealed envelopes, each bearing on the outside only the name of the hospital and a number. After acceptance of a patient by the panel, and before admission to the streptomycin trial, the appropriate numbered envelope was opened at the central office; the card inside indicated whether the patient was to be an S or a C case, and this information was then given to the medical officer of the centre.

Subsequent analysis showed that random allocation had led to the two groups being comparable at entry to the trial. After six months, 51 percent of the treated group showed considerable radiological improvement (radiographs were assessed without knowledge of the treatment group); only eight percent of the control group showed such improvement. Seven percent of the treated group were dead in six months as opposed to 27 percent of the control group.

The ethical considerations did not present a problem as bed rest was considered to be the only possible alternative treatment and only limited supplies of streptomycin were available. As not all cases could be given the new drug, it was reasonable and practicable to give it to a random half. The randomised controlled trial was therefore born nearly 50 years ago and has since gone from strength to strength.

The most recent innovation has been for trials collecting the minimum of information on outcome but in a very large number of subjects. The results of these major trials, for example the ISIS trials, have led to large changes in clinical practice [31].

3.   ETHICAL CONSIDERATIONS

DEFINITION

The *Concise Oxford Dictionary* defines ethics as the *"science of morals."* Glaser considered that a discussion of ethical problems should embrace both an assumption of right and wrong and a definition of how things are and not just how things should be [32]. Moreover, ethical problems concern the individual rather than the community. The community may benefit from the results of a trial but no individual should be asked to take an unreasonable risk to benefit the community. Problems arise when we are forced to consider what is reasonable.

Ethical considerations are not legal requirements, but the law may support an ethical stance. Lawyers usually consider precedents and determine the truth of matters by discussion. We can emulate this process for a definition of the term *reasonable*. At one extreme, it is obviously not ethical to force (or even request) a subject to take part in a dangerous study. Such trials were performed on non-Aryan prisoners in Nazi Germany. Subjects were exposed to extremes of temperature and trials of resuscitative techniques capable of saving the lives of many in the community, these trials were obviously unethical and the risks to the individual unreasonable. At the other extreme, every patient who agrees to take a medication must accept some risk. There are risks to taking penicillin or aspirin but most subjects would be willing to face these risks in a controlled trial. This is an example of a reasonable or acceptable risk.

LEGAL CONSIDERATIONS

As Wade has pointed out [33]. *"Although the subject needs protection, the community needs knowledge."* He considered how a subject should be indemnified if matters go wrong. The institution where the trial takes place must have a public liability insurance policy in case anything untoward happens to a subject as a result of negligence. With a new drug not in ordinary use, the policy may not cover such a contingency and, where applicable, the pharmaceutical company should agree to carry the risk. We support Wade's idea that institutions should have no fault liability insurance so that subjects in trials may claim compensation for injury even when negligence does not occur. For example, a patient who experiences an adverse drug reaction while taking part in a trial could be recompensed.

DECLARATIONS ON MEDICAL ETHICS

The Nuremberg Code

Following trials of Nazi war criminals, ten standards were laid down in 1947 [34].

1. The subject must give his or her voluntary consent, knowing the nature, direction, purpose, inconveniences, and hazards of the experiment.
2. The experiment should be necessary both in yielding fruitful results for the good of society and in the sense that the information cannot be gained without experiment.
3. The anticipated results justify doing the experiment (see sections in the declaration of Helsinki: Clinical Research Combined with Professional Care, and Nontherapeutic Clinical Research).
4. All unnecessary physical and mental suffering must be avoided (see section on The Use of Sham Operations).
5. There should be no prior reason to believe that death or injury will occur.
6. The degree of risk shall not exceed the humanitarian importance of the problem.
7. Preparations should be made and adequate facilities provided against the remote

possibility of adverse effects.
8. Those who conduct the experiment shall exercise the highest degree of skill and care and be scientifically qualified.
9. The subject must always be free to bring the experiment to an end.
10. The investigator must terminate the experiment if its continuation may be detrimental to the patient.

The declaration of Helsinki

The World Medical Association produced the following declaration [35], prefaced by binding the doctor with the words, *"the health of my patient will be my first consideration."*

I. *Basic Principles*
   1. Clinical research must conform to the moral and scientific principles that justify research, and should be based on laboratory and animal experiments or other scientifically established facts.
   2. Clinical research should be conducted only by scientifically qualified persons and under the supervision of a qualified medical man.
   3. Clinical research cannot legitimately be carried out unless the importance of the objective is in proportion to the inherent risk to the subject.
   4. Every clinical research project should be preceded by careful assessment of inherent risks in comparison to foreseeable benefits to the subject or to others.
   5. Special caution should be exercised by the doctor in performing clinical research in which the personality of the patient is liable to be altered by drugs or experimental procedure.
II. *Clinical Research Combined with Professional Care*
   1. In the treatment of the sick person the doctor must be free to use a new therapeutic measure if in his judgement it offers hope of saving life, re-establishing health, or alleviating suffering.

If at all possible, consistent with patient psychology, the doctor should obtain the patient's freely given consent after the patient has been given a full explanation. In case of legal incapacity consent should also be procured from the legal guardian; in case of physical incapacity the permission of the legal guardian replaces that of the patient.

2. The doctor can combine clinical research with professional care, the objective being the acquisition of new medical knowledge, only to the extent that clinical research is justified by its therapeutic value for the patient.

III. *Nontherapeutic Clinical Research*

1. In the purely scientific application of clinical research carried out on a human being it is the duty of the doctor to remain the protector of the life and health of that person on whom clinical research is being carried out.

2. The nature, the purpose, and the risk of clinical research must be explained to the subject by the doctor.

3a. Clinical research on a human being cannot be undertaken without his free consent, after he has been fully informed; if he is legally incompetent the consent of the legal guardian should be procured.

3b. The subject of clinical research should be in such a mental, physical, and legal state as to be able to exercise fully his power of choice.

3c. Consent should as a rule be obtained in writing. However, the responsibility for clinical research always remains with the research worker; it never falls on the subject, even after consent is obtained.

4a. The investigator must respect the right of each individual to safeguard his personal integrity, especially if the subject is in a dependent relationship to the investigator.

4b. At any time during the course of clinical research the subject or his guardian should be free to withdraw permission for research to be continued. The investigator or the investigating team should discontinue the research if in his or their judgement it may, if continued, be harmful to the individual.

The Helsinki declaration clearly differentiated between the situation when the subject, usually a patient, can hope to benefit from the experiment and the situation where no such benefit can be expected.

Section II. 2 of the declaration stated that clinical research can be combined with professional care *"only to the extent that clinical research is justified by its therapeutic value for the patient."* This must be the overriding ethical consideration and the use of patients as volunteers for experiments not relevant to treatment presents great difficulties and will be discussed in a later section.

Sir Austin Bradford Hill has taken issue with two recommendations of the World Medical Association [36]. He found that there are experiments such as *"in industrial psychology - which are not the prerogative, or even within the special competence, of the medically qualified,"* and he therefore objected to item I. 2, which insisted on the supervision of a qualified medical man. Hill also disagreed with the idea that the nature and purpose of the trial must be explained to the subject and stated *"... I have no doubt whatever that there are circumstances in which the patient's consent to taking part in a controlled trial should be sought. I have equally no doubt that there are circumstances in which it need not - and even should not - be sought."*

The necessity, or otherwise, for informed consent has remained controversial. Glover emphasised the need for informed consent except in the unconscious or those unable to understand [37]. In response to this article, King, replying on behalf of patients stated *"we must also recognise that there are costs (as well as benefits) of 'fully informed consent' - namely fear and uncertainty"* [38]. Simes and colleagues [39], when conducting a trial of treatment for cancer also randomised patients to two methods of seeking consent: one limited to the aims, expected results and potential toxicities of treatment and the second requiring disclosure of all relevant information including the fact that the patient had cancer, treatment was part of a research study and allocated randomly and the existence of alternative treatment. Patients receiving total disclosure tended to have a better understanding but to be more anxious and less willing to agree to randomisation. Three to four weeks later those differences had disappeared.

Obtaining informed consent requires combining the roles of scientific researcher and caring physician. This difficulty has led to doctors agreeing to take part in a trial and then failing to enrol patients. Apparently *"the clinician is confronted with the uncongenial task of confessing to both ignorance and impotence"* [40].

Bradford Hill's specific questions

Bradford Hill was unhappy with codes that deal in generalities and take no heed of *"the enormously varying circumstances of clinical medicine"* [36]. He stressed the necessity for *"The close and careful consideration in the specific circumstances of each proposed trial"* and formulated a series of questions to be answered for each trial.

1. Is the proposed treatment safe or, in other words, is it unlikely to do harm to the patient?
2. Can a new treatment ethically be withheld from any patients in the doctor's care? Tuberculous meningitis was a universally rapidly fatal condition and when the first case reports revealed that streptomycin treatment had resulted in the patients' recovering, this fact was conclusive evidence of the effectiveness of the new treatment. It was then not ethical to perform a clinical trial of streptomycin in tuberculous meningitis. However, respiratory tuberculosis runs a more variable course and it was ethical to perform the randomised controlled trial of streptomycin in this

condition. Moreover, only a limited amount of streptomycin was available at that time (1947) and as all cases could not be treated, it can be argued that it would be unethical not to have performed the trial.

3. What patients may be brought into a controlled trial and allocated randomly to different treatments?
4. Is it necessary to obtain the patient's consent to his inclusion in a controlled trial?
5. Is it ethical to use a placebo or dummy treatment?
6. Is it proper for the doctor not to know the treatment being administered to his patient?

Medical Research Council

A statement by the Medical Research Council (MRC) [41] gave two examples of when informed consent may not always be desirable. For example, when the patient has a possibly fatal illness without effective treatment being available, and second, when a placebo is employed. The MRC considered in 1964 whether any supervision of the conduct of controlled trials (or other experiments) was necessary and concluded *"controlled clinical trials should always be planned and supervised by a group of investigators and never by an individual alone."* The MRC report also suggested that no paper should be accepted for publication if there are any doubts about the ethical conduct of the study leading to the report. In certain trials funded by the MRC the issue of informed consent has not been handled clearly and has led to criticism [37,42].

RESEARCH ETHICAL COMMITTEES

In 1967 a committee appointed by the Royal College of Physicians of London suggested the formation of ethical committees consisting of *"a group of doctors including those experienced in clinical investigation"* [43]. By 1973 the functions and constitution of these committees had been formalised.

The final report made the following recommendations:

1. A Research Ethical Committee shall be a small committee set up solely to supervise the ethics of clinical research.
2. The medical members should be experienced clinicians with knowledge and experience of clinical research.
3. The Research Ethical Committee should have a lay member.
4. To remove any uncertainty about which procedures should be submitted to a Research Ethical Committee, all proposed research investigation in human beings should be submitted.
5. Whenever a research investigation was not expected or intended to benefit the individual patient a full explanation should be given and the patient should be free to decline to participate or to withdraw at any stage.
6. Whenever possible the consent of a patient should be obtained in the presence of a witness.
7. When there are circumstances in which it is genuinely inappropriate to inform a patient fully, it is the duty of the Research Ethical Committee to examine the situation with special care.
8. Particular care is needed if clinical investigation is proposed in children or mentally handicapped adults who cannot give informed consent. The parents or guardian should be consulted.
9. Particular care is needed if clinical investigation is proposed on a subject or patient who has any sort of dependent relationship to the investigator, for example, student, laboratory technician, or employee.

The importance of the ethical committee is without doubt but whether numerous local committees need to consider a multicentre protocol has been questioned. Wald stated *"Clearly, the ethical aspects of each preventive trial need consideration, but this should be the responsibility of a central ethical committee and not a matter to be considered separately by many (perhaps over a 100) district research ethical committees that are, in the main, concerned with local clinical research projects. The view that the ethical aspects of preventive trials can only be properly considered locally is unjustified. In general, the opposite is the case: the scale of the research and the kind of expertise required to assess the scientific, and therefore the ethical, validity are such that the research is best considered centrally"* [44]. A legal requirement for a local committee to examine a proposal may be overcome by the taking of 'chairman's action'.

RANDOMISED CONTROLLED TRIALS WITHOUT POSSIBLE BENEFIT TO THE PARTICIPANT

Examples of these trials are provided by early drug studies in normal men and women and trials of drug interactions in patients on chronic treatment. Early drug trials in normal men are usually dose-finding experiments to assess the human counterpart of observations made in animals. They are not, initially, randomised controlled trials but slightly later studies may constitute a randomised trial of the new treatment (in the predetermined dose) versus an established drug (chapter 18).

In trials on patients, those on chronic treatment with one drug may be asked to take a second drug to assess the effect of the drugs in combination. This may be suggested when the second drug cannot be expected to benefit the patient. An example can be given of patients on long-term antihypertensive drugs who are also asked to take an antidepressant or antiinflammatory drug to assess whether or not the second drug worsens blood-pressure control.

For trials without possible therapeutic benefit for the individual, all subjects and patients must be true *volunteers*, receive *full information* about the study, give *written consent* (preferably in front of a witness), and *not receive an excessive reward.* If the subjects are paid a considerable amount they may be tempted to participate in a study, whereas without this remuneration they may refuse. This restriction does not exclude an allowance for fares, meals, and compensation for lost earnings as volunteers should not be expected to experience a financial loss.

Who should participate in trials without possible therapeutic benefit?

Employees of the pharmaceutical industry

Glaser outlined the case very clearly for using employees of the pharmaceutical industry [32].

Those who decide that a new substance can be safely tried in man should have enough confidence to take it themselves. If they will not take it themselves, they should not give it to others. Those who know the most about the substance and who are the most experienced scientists can make the best personal decisions about it and they are also the best able to observe their own subjective effects. Thus the first to take a new substance might be the research director, the medical director, the senior toxicologist, or advisers in pathology.

Glaser also considered that a volunteer's family doctor should be informed about the trial. This may ensure that trials that are unacceptable to general practitioners are not performed, and if there is some medical reason why an individual volunteer should not participate, then the investigator may be informed of this fact. Lastly, if the volunteer should become ill during or after the trial, then the general practitioner will be aware that the trial is in progress.

Volunteers must not be solicited from subordinates by their seniors. Only a comparatively junior person should perform this task and the supervisors should be told only who is suitable. Glaser reported *"anyone unwilling is unsuitable"* and children under the age of 14 and mental patients cannot volunteer.

The alternative view has been expressed by Hamblin "... *employees of the firm or institution doing the experiments are not suitable subjects*" [45]. This remark followed a review of experiments performed of "*radioactivity simply to satisfy scientific curiosity*" and Hamblin is probably in a small minority when he states "*Trials in healthy subjects must always be ethically dubious ..*". However most would agree that these studies "*are a breach of human rights if participants are not fully informed or are coerced into taking part*".

Prisoners

Prisoners are used in medical experiments in the United States of America. The problem with this procedure is that a reduction in prison sentence may constitute an excessive reward and result in the subjects not being free volunteers. The report of a Committee appointed by the Governor of Illinois stated [46]:

A reduction in sentence in prison, if excessive or drastic, can amount to undue influence. If sole motive of the prisoner is to contribute to human welfare, any reduction in sentence would be a reward. If the sole motive of the prisoner is to obtain a reduction in sentence, an excessive reduction of sentence which would exercise undue influence in obtaining the consent of prisoners to serve as subjects would be inconsistent with the principle of voluntary participation.

The committee considered the function of imprisonment, for example, whether this is to protect society or to reform the prisoner. The members discussed whether a prisoner would volunteer from good social consciousness or in a desire to reduce his sentence. In view of the latter incentive, the committee concluded that a prisoner should not be allowed to volunteer if he is a habitual criminal or if he has committed a notorious or heinous crime. Presumably, the committee members were worried about having such a person released early.

The committee also concluded that any proposed reduction in sentence must not be excessive. Glaser [32] also worried that the incentives for prisoners may be too high. He considered the possibility of prisoners getting privileges for participation and even that the relief of boredom might prove a great incentive, possibly a coercion inconsistent with voluntary participation.

Patients

Patients are the ultimate beneficiaries of advances in medical care and Claude Bernard considered it their duty to assist with research. However, should the individual patient in the trial be the possible beneficiary or should the benefit go to other patients with different conditions? If we wish to assess the interaction between an antihypertensive and antiinflammatory drug, we may ask any hypertensive patient or patients with both hypertension and arthritis to cooperate. In the latter instance, the treatment is relevant to the patient's condition and the patient may benefit. However, when the patient has hypertension alone, he must be considered as a normal volunteer and great care must be taken that the doctor-patient relationship is not used to exert too much pressure on the patient to participate. The patient must not volunteer from a sense of gratitude or in the hope of better medical attention and it is a wise precaution for the doctor treating the patients to ask them to discuss taking part in a trial with another colleague. The doctor undertaking the usual treatment should make it clear that the patient's failure to participate in a trial will not affect his usual medical care in any way.

INFORMED CONSENT

Information for the patient or subject

The patient or subject should be fully informed of the nature of the trial: that is, the number of investigations and visits required and the duration of the trial. The objectives of the trial should be stated, provided such statements are compatible with the usual doctor-patient relationship. As discussed above it may be unethical to give full information to patients with, say, cancer, either when the diagnosis cannot be revealed or when it is not in the patients' best interest to describe the inadequacies of available treatment. However, in most instances the patients can be given all the relevant information and should be told when a placebo (dummy) treatment is to be employed in the trial. In conclusion, the patients should be informed of the following, in writing, and preferably in the presence of a witness:

1.   the nature of the treatments being compared
2.   the objectives of the trial
3.   the duration of the trial
4.   what the trial involves for the patient (number of visits, investigations, et cetera)
5.   possible benefits to be derived from the treatments
6.   possible hazards of the treatments
7.   what to do if the patient becomes unwell, runs out of tablets, et cetera
8.   alternative treatments
9.   the fact that the patient may withdraw from the trial at any time
10. the fact that not entering the trial will not jeopardise usual medical care

Written consent

Written consent should be obtained; otherwise, there can be no proof that consent was given and such evidence may be necessary in a court of law. The patients should be asked to sign a document giving the full information discussed above and including a declaration similar to the following: *"I ......, have read the above description of the trial and agree to take part. I understand that I may withdraw my cooperation at any stage should I so wish."* The patients therefore sign to say they have been informed about the trial and have agreed to take part. Some authorities may insist that the declaration be signed in the presence of a witness. This is desirable in all volunteer studies but many researchers would not insist on the presence of a witness when the trial is of possible therapeutic benefit to the individual involved.

How to avoid asking consent of some of the patients

When a new experimental treatment is to be compared with an acceptable routine treatment it can be argued that only the patients receiving the new treatment need give consent. The usual trial design requires consent to be obtained prior to randomisation but Zelen has suggested that randomisation can precede informed consent so that only those allocated to the new treatment are asked to consent [47]. There may well be a circumstance in which this strategy is desirable; however, the approach is impossible in double-blind or single-blind trials. Moreover, Zelen's suggestion may be unsatisfactory if the new treatment proves to represent an important new advance: the patients who have benefitted will have consented to take part but not those who have fared badly. Most important, however, is the fact that those who do not wish to take part will fail to receive the trial treatment in the consent group but will receive it in the no consent group. Patients who do not receive the treatment cannot be excluded; otherwise, the two groups may be dissimilar for important characteristics and one major purpose of randomisation will be lost. Analysis has to be conducted on all randomised patients on the intention-to-treat principle (chapter 14). However, the effect of the new treatment may be diluted by the results in patients who do not receive this treatment.

Poor recruitment occurred in the North American National Surgical Adjuvant Project for Breast and Bowel Cancer (NSABP) trial to compare total mastectomy with segmental mastectomy in the treatment of breast cancer [48]. The Zelen model was adopted and the recruitment rate increased sixfold. the scientific price was the increased risk of systematic bias and the Cancer Research Council Working party rejected the Zelen model

[49]. This working party considered that failure to get informed consent could lead to an action for battery or negligence and considered that informed consent should be obtained in two stages, first an agreement in principle and second a later signing of a consent form. Ideally the working party considered that a trained counsellor should assist.

PLACEBO TREATMENT

The reasons for using placebo tablets and the methods for using such tablets are discussed in chapter 7. There are two circumstances where it is ethical to employ placebo medication; when no effective treatment is available for a particular condition and when, if such treatment is available, it can safely be withheld for a certain period.

No effective treatment has been identified

A placebo cannot be employed if there is definite evidence that withholding standard treatment would be detrimental to the patient's health. Beecher [50] discussed a trial of treating streptococcal respiratory infections in which placebo was given to 109 men while benzanthine penicillin G was given to the others. No patient treated with penicillin developed either rheumatic fever or acute nephritis. However, three patients developed these complications when given placebo. Beecher considered that at the time the trial was performed it was known that penicillin prevented rheumatic fever, and therefore the use of a placebo was unethical.

Use of a placebo for a short period when active treatment is known to be required

We can agree that it is unethical to withhold necessary treatment. However, a placebo may have a powerful pain-relieving effect and constitute acceptable treatment under certain circumstances. Beecher [51] reviewed the effects of placebo in severe postoperative wound pain. Four studies used an injection of saline as a placebo and satisfactory pain relief was achieved in about a third of patients. Similarly, placebo treatment produced relief from angina pectoris in a similar proportion of patients.

In patients with severe postoperative pain, we can argue that placebos should not be used, as active drugs such as morphine are available. But arguments for using a placebo in this situation can be advanced. A proportion of patients achieve pain relief from placebo and they do so without the adverse effects associated with active pharmacological agents. Although it would be unethical to withhold an active analgesic for a prolonged period, administration may be delayed for a short interval, say, 15-30 minutes following a placebo injection. If pain relief is not achieved at the end of this period, active treatment can then be given. Many patients will agree to wait a short period to help in evaluating a new drug. Care has to be taken in a double-blind study of a new drug against placebo that there is no likelihood of an adverse effect if the new drug is ineffective and is closely followed by an active drug such as morphine.

With less severe degrees of pain the ethical problems are reduced and it is even more necessary to employ a placebo to assess a possibly active mild analgesic. Many patients will respond adequately to a placebo and not all patients receiving active treatment will experience pain relief. For example, if the improvement rate is 33 percent with placebo and 50 percent with an active treatment, the use of placebo is necessary to confirm a superior effect for the active compound. Without knowledge of the placebo response rate, a 33 percent result could be a nonspecific response to an inactive compound.

Placebo treatment has also been employed for short periods in trials in chronic diseases when active treatment is known to be required in the long term. An example is provided in hypertension where placebo treatment in the long term is justified for patients with mild hypertension as the benefits of active treatment have not been established. However, active treatment is known to be beneficial in preventing cerebrovascular events in young or middle-aged patients with moderate or severe hypertension. Yet placebos are prescribed when immediate treatment for heart failure, renal failure, or malignant hypertension is not required. Placebo treatment is traditionally employed in two broad circumstances.

1. *The first is when the patient has not received antihypertensive treatment in the past.* Antihypertensive treatment is not usually started the first time the patient sees a doctor. The physician may wish to confirm that the blood pressure is elevated on a second or third occasion and may require certain investigations to be completed before commencing treatment. It is therefore reasonable to give placebo treatment during this period of observation and possibly to extend the interval to, say, four to six weeks. It must be appreciated, however, that the preventive effects of antihypertensive treatment are being denied the patient during this period. Although the risk of a cerebrovascular event occurring during a short interval is low, it still constitutes an ethical problem. The theoretical risk for moderate/severe hypertension can be calculated from the Veterans Administration trial of antihypertensive treatment [52]. Male patients less than 60 years old with an initial diastolic blood pressure of 105-114 mmHg experienced, on average, a 0.058 chance of a mortal or morbid cardiovascular event per year of placebo treatment. If they were given active treatment the chance of a cardiovascular event was 0.023 per year. The excess risk of being on a placebo was therefore 0.035 events per year and the probability of a mishap with placebo treatment may be 0.003 per month. If 220 such patients receive six weeks of placebo treatment in a trial, the investigator may expect one adverse event. These events include stroke, myocardial infarction, heart failure, and the retinal changes of accelerated hypertension. Trials have subjected patients to a one in 220 risk of one of these events. However, placebo

treatment during run in is now usually applied in trials studying patients with mild/moderate hypertension and consequently, lower risks.

2. *The second circumstance is when placebo treatment interrupts a period of active treatment.* The risks of taking placebo treatment may be the same when a period of placebo treatment interrupts active treatment as when it precedes it, and many patients on treatment have been entered into trials that incorporate a period of placebo administration. Again, for a six-week period we must ask if the patient will accept a one in 220 chance of an adverse cardiovascular event. Wash-out periods in cross over trials (chapter 5) usually only last one or two weeks.

In conditions other than hypertension, the use of placebos is more contentious. Anxiety has been expressed that any relapse of schizophrenia may have long term adverse consequences. In a nine month trial of fluphenozine decanoate for schizophrenia, 8% relapsed on active treatment and 66% on placebo. Seven years later, however, there was no difference between the groups and no evidence of relapse producing a long term deterioration [53].

Following discussions with patients with AIDS and their advocates, Rebecca Pringle Smith is quoted as saying *"Even if you have a supply of compliant martyrs, trials must have some ethical validity"* [54]. With hindsight, in 1995 the AIDS patients given placebo were not adversely affected compared with active treatment and were not martyrs.

The use of sham operations

The immediate reaction among physicians is to consider all sham operations in man as unethical. Bradford Hill considered that it would not have been reasonable to use placebo injections as a control in the Medical Research Council (MRC) trial of streptomycin [5]. In this trial the control patients would have suffered a considerable amount of discomfort from repeated injections of placebo. Admittedly, if injections alone can have a life-prolonging effect independent of the substance injected, then the MRC trial was biased in favour of streptomycin. However, such a strong placebo effect was unlikely and could not justify so much discomfort on the control patients.

Before dismissing sham operations, however, we must consider a trial discussed by Beecher [55]:

In 1939 it was suggested, in Italy, that the pain of angina pectoris could be greatly lessened by ligation of the internal mammary arteries. Eventually this suggestion was adopted in the United States and quite spectacularly favourable results were obtained. Not only were the objective results impressive, the patients said they felt better and the objective evidence supported this: there was great reduction in the number of nitroglycerin tablets taken, and exercise tolerance was greatly increased. Several individuals [56-59] began to wonder if this might not be a placebo effect. They therefore went to their patients, explained the situation, and told them they would like to carry out a study in which the patients would not know what had been done, nor would the observers know until the study was completed. They told their patients that half of them would have the internal mammary arteries exposed and ligated and the other half would simply have them exposed, but not ligated. These studied were carried out ... ligation had no real effect beyond that of a placebo effect.

Beecher thus argued very persuasively that a sham operation can be ethical even though the control patients suffered an anaesthetic and much discomfort. Many patients would not agree to take part in such a trial.

SELECTION (EXCLUSION OR INCLUSION) CRITERIA

Exclusion and inclusion criteria are the two sides of the same coin. A trial confined to young patients may be said to *exclude* patients above the age of 60 or *include* only patients below the age of 60.

Two objectives are met by using these criteria. First, only those patients who are intended for study are entered into the trial. The results of the trial are then only valid for a similar group of patients; this concept will be discussed further in chapter 13. The second reason for having selection criteria is ethical. Patients must be excluded from a trial if inclusion in the trial may produce adverse consequences for them.

Table 3-1 gives the selection criteria for a placebo-controlled trial of antihypertensive treatment in the elderly, conducted by the European Working Party on Hypertension in the Elderly (EWPHE) [60]. The selection criteria are rearranged into criteria defining the group of patients to be studied, and criteria excluding patients from the study who should not be included for ethical reasons. The trial involved the random allocation of patients either to over five years of active treatment or over five years of placebo treatment. The selection criteria therefore exclude patients who should not receive a placebo for this period. At the time of initiating the trial, there was little or no evidence that the elderly hypertensive patient would benefit from antihypertensive treatment. However, it was considered desirable to exclude patients with very high levels of blood pressure (criteria 9-10). Similarly, hypertensive patients known to require treatment were excluded, for example, those with accelerated or malignant hypertension (criterion 11), those with congestive heart failure, and those with conditions that would possibly benefit from treatment, such as patients with renal impairment (criterion 13) or those who had previously suffered a haemorrhagic stroke (criterion 14).

Selection criteria must exclude not only those patients who would suffer from placebo treatment, but also those known to be adversely affected by the active treatment. Criteria 16-17 exclude patients who may be adversely affected by the active treatments employed in the trial (hydrochlorothiazide with triamterene; and methyldopa).

Table 3-1. Selection Criteria for the European Working Party on High Blood Pressure in the Elderly (EWPHE) Trial of Antihypertensive Treatment.

Selection criteria defining the group of patients to be studied

1. Aged more than 60.
2. Systolic blood pressure (on placebo) above 160 mmHg.
3. Diastolic blood pressure (on placebo) above 90 mmHg.
4. Patients give their informed consent.
5. Regular follow-up possible.
6. Compliant with medication as assessed by pill count.
7. No reason to suspect secondary hypertension.
8. No severe life-threatening diseases unrelated to hypertension (eg. carcinoma).

Selection criteria included for ethical reasons

9. Systolic blood pressure (on placebo) not above 239 mmHg.
10. Diastolic blood pressure (on placebo) not above 119 mmHg.
11. No history of accelerated or malignant hypertension.
12. No congestive heart failure
13. No severe renal failure (serum creatinine >2.5 mg %).
14. No previous history of a haemorrhagic stroke or hypertensive encephalopathy.
15. No history of dissecting aneurysm.
16. No previous history of gout or serum uric acid >10 mg %.
17. No acute hepatitis or active cirrhosis.

WITHDRAWALS FROM THE TRIAL

If a patient with a certain condition cannot enter the trial for ethical reasons, then he should be withdrawn from the trial if he develops the condition. Withdrawal criteria should therefore be the same as exclusion criteria. Table 3-2 gives the withdrawal criteria for the EWPHE trial. Criteria 1-3 are end points for the trial and not ethical considerations. Criteria 4-10 correlate with the exclusion criteria given in table 3-1. The criteria in the two tables are cross-referenced in table 3-2. Selection criteria 16-17 (table 3-1) do not have their counterparts in the withdrawal criteria in table 3-2, as the development of gout may lead to the discontinuation of diuretic treatment, but the patient may continue in the trial taking methyldopa. Similarly, the development of liver disease may lead to stopping methyldopa, with the patient continuing to take a diuretic and remaining in the trial. Withdrawal criteria 11 and 12 have no counterpart in the selection criteria but indicate that the patient is not progressing satisfactorily.

However carefully a trial is designed, and even after the completion of a pilot trial, there will still be patients whose continuation in the trial would be against their future well-being. Criterion 13 allows for these unforeseen contingencies and is a necessary statement in any trial.

Table 3-2. Withdrawal Criteria for the European Working Party on High Blood Pressure in the Elderly (EWPHE) trial.

Withdrawal criteria which are end-points for the 'on-randomised treatment' part of the trial and not ethical considerations.

1. Completion of agreed period of follow-up.
2. No follow-up for more than six months.
3. No trial treatment for more than three months.

Withdrawal criteria for ethical reasons from the 'on-randomised treatment' part of the trial.
(In parentheses - the corresponding numbers for the selection criteria)

4. Systolic blood pressure rising by 40 mmHg or exceeding 250 mmHg on three visits (9).
5. Diastolic blood pressure rising by 20 mmHg or exceeding 130 mmHg on three visits (10).
6. Development of accelerated or malignant hypertension (11).
7. Development of congestive heart failure (12).
8. Serum creatinine increasing by 100% or above 3.9 mg % on two occasions (13).
9. Development of cerebral or subarachnoid haemorrhage or hypertensive encephalopathy (14).
10. Development of a dissecting aneurysm (15).
11. Voltage criteria for a 30% increase in left ventricular hypertrophy as assessed from the electrocardiogram.
12. A 20% increase in the cardio-thoracic ratio as measured on a chest radiograph.
13. Any reason why continuation in the trial would be detrimental to the patient's interest.

STOPPING THE WHOLE TRIAL

Decision rules for stopping are considered in chapter 12. The present section deals with ethical aspects. In a short-term trial, the patients are usually entered into the trial quickly and the trial completed before the results are analysed. The exception to this rule is the sequential trial where a decision is made whether or not to continue with the trial as the individual results become available (chapter 5). When patients are followed for several years or when recruitment persists for many years, the opportunity exists for interim analyses to be made. For ethical reasons the trial must be terminated if an interim analysis demonstrates a statistically significant and important adverse effect of treatment or a significant benefit from treatment. If interim analyses fail to reach these end points the trial will be terminated when the intended number of patients has entered the trial or the trial participants and organisers run out of time or money (chapter 12).

Problems with significance testing in interim analyses

Care has to be taken that the overall level of significance of a trial is not reduced by the repeated analysis of results. These interim analyses are sometimes termed *repeated looks*. If a statistical test is repeated on several occasions on increasing data and five percent is taken as the level of significance to be achieved, then after the first test the probability of a falsely positive result is five percent. After two tests this probability rises to almost ten percent. After 13 tests, the chances of a falsely positive result is almost 50 percent.

McPherson [61] has calculated that ten interim analyses with a decision rule to stop the trial if the level of significance exceeds one percent is equivalent to an overall level of significance of five percent (chapter 12). In other words, if the trial must be stopped for an adverse effect significant at the five percent level and ten interim analyses are planned, then the result of an interim analysis must be significant at the one percent level to stop the trial.

Terminating the trial when an adverse effect of treatment is observed

Treatment with either conjugated oestrogens or dextrothyroxine in the Coronary Drug Project Trial [62] had to be terminated owing to the adverse effects of these drugs. Similarly the University Group Diabetes Project (UGDP) trial was stopped owing to the adverse effects of phenformin and tolbutamide [63] (chapter 19). In both trials the groups treated with certain drugs fared significantly worse than the placebo-treated groups and the trials of these active treatments were terminated.

Terminating the trial when a statistically significant benefit is observed

The Veterans Administration trial of antihypertensive medication provides a good example of a trial's being terminated when an interim analysis provides evidence of a benefit from treatment. Patients were entered into the trial when the diastolic blood pressure ranged from 90-124 mmHg while taking a placebo and they were randomly allocated to receive either active or placebo treatment. After an average of 18 months' follow-up the trial was stopped for patients with an initial diastolic blood pressure greater than 114 mmHg [64], as the patients receiving active treatment had fewer cardiovascular events than those receiving placebo ($P<0.001$). The trial was continued for patients with an initial diastolic blood pressure of 90-114 mmHg [52]. Another interesting example comes from the Anturane Reinfarction Trial [65] (chapter 19). In this trial a significant benefit from treatment was observed in an interim analysis and recruitment to the trial was stopped. However, the patients already in this double-blind trial were advised of the results of the interim analysis and asked to continue the trial. Nearly all suggested to continue and the final analysis showed a similar benefit.

It has been agreed that the interim findings should be presented to investigators and patients to ensure that no one receives what may be an inferior treatment [66]. However, a demonstration of 'possible trends' may lead to physicians stopping randomising patients and patients withdrawing from the study. No proof would have been obtained as to the best treatment and yet the trial may stop. Ethical problems will arise if the trial result (or lack of it) is not accepted, not least as the participation of the patients will have been wasted and the trial needs to be started again in the future. The results of interim analyses should not be communicated either to the investigator or the patients, although they should know the conditions under which the trial would be stopped.

TRIALS TO DETECT TOXICITY

It is unethical to design a trial to detect toxicity. However, as discussed in chapter 16, large long-term trials have resulted in the detection of unexpected treatment toxicity, although some large trials have failed to detect rare adverse drug effects. Large trials are designed to estimate the efficacy of treatment but careful attention must be paid to possible toxicity.

When may a trial to detect both efficacy and toxicity be desirable?

When a single trial reports a benefit from treatment, it is often desirable to repeat the trial and ensure that the benefits can be demonstrated for different patients and on another occasion. However, when a trial detects toxicity, it is ethically impossible, although scientifically desirable, to conduct a trial to confirm an adverse effect of treatment. If there are doubts about whether or not there was any serious toxicity in the earlier trial, and the efficacy of the treatment is thought to be high, then possibly a second trial of benefit can be mounted. The question is of some practical importance. For example, in the University Group Diabetes Programme Trial [63] oral hypoglycaemic agents were associated with an increase in cardiovascular mortality. However, when a patient cannot adhere to a diet these drugs may relieve the symptoms of hyperglycaemia. It may be reasonable to reassess efficacy in these patients and in view of the criticism levelled at this particular trial (chapter 19), the trial could be repeated. I would be reluctant to take part in a trial where toxicity may be a disadvantage *not* counteracted by important gains from therapy.

THE INFLUENCE OF FINANCIAL REWARDS

The financial rewards to the investigator or the volunteer must not be excessive. Both parties must avoid unreasonable risks and risk-taking would be increased for large financial gains. Waldron and Cookson [67] have pointed out that the name of a trial sponsor may not be known (the trial being conducted through a contract research organization), and that the results of the research are often not published. Most research is therefore done for financial and not scientific advance. The British Medical Association recommends £109.50 (1993) for one hour of a doctor's time in a clinical trial. Ethics committees can require a payment schedule to be attached to the protocol for approval [68]. These rulings may limit the amount of research done for commercial reasons but the underreporting of research remains a problem. Herxheimer and Chalmers [69,70] consider this to be *"a form of scientific misconduct because it distorts the publicly available evidence."* This waste of resources could be rectified by publishing results in electronic journals with unlimited space [69] and by registering all trials at inception[71].

An additional problem is when investigators are aware of promising results and are capable of profitable but illegal share dealing. The organisers of one study decreed that *"clinical investigators and their spouses and dependents should not buy, sell, or hold stock or stock options in any company providing or distributing medicines under study"* [72].

CONCLUSIONS

This chapter reviews the ethical requirements in the design and conduct of clinical trials. Declarations of ethical principles have been reviewed and the place of research ethical committees considered. Emphasis was placed on the importance of obtaining informed written consent and a distinction has been drawn between trials of possible benefit to the participant and trials involving volunteers who cannot expect an improvement in their health from participating in the trial. The investigator must remain convinced that none of the available treatments offer a clear advantage and this is especially important when placebo treatment is to be employed. Provided the investigator is genuinely in doubt as to the best treatment, he can explain the situation to potential participants and ask them to enter the trial. He may even ask himself the standard question, *"Would I allow a member of my family to enter the trial?"* Even if he can answer yes to this question, the public must be protected from a small proportion of eccentric enthusiasts; research ethical committees should provide this safeguard.

Large trials should incorporate an ethical committee in the administration that is independent of the investigators and rules on whether any observed toxicity is acceptable, when the trial should be terminated, and whether to make any changes in the criteria for entry or withdrawal from the trial.

If a trial shows one treatment to be superior, patients who received inferior treatment may have suffered as a consequence. Trial designs that limit this problem are discussed in chapter 5 and the ethical disadvantages of randomised controlled trials are summarised in chapter 20.

4. THE OBJECTIVES OF A RANDOMISED CONTROLLED TRIAL

A trial may be conceived to test more than one hypothesis but it is good practice to determine one or a limited set of major objectives. For example, an investigator may be interested in a trial of a new antihypertensive drug in elderly patients. The major objective could be either to demonstrate the efficacy of the drug in lowering blood pressure or in preventing cardiovascular deaths. The first objective could be answered in a few patients studied for six months, but the second objective would require the study of hundreds of patients over many years (chapter 6). In order to calculate the numbers required for a trial, the major objectives have to be identified and the smallest treatment effects to be detected must be defined.

IDENTIFICATION OF THE MAJOR OBJECTIVES

The major objectives will involve the detection of a change in particular end points. In the above example, where the effect on blood pressure has to be determined, the main end points of interest could be diastolic, systolic, or mean pressure. If the effect on mortality or morbidity has to be determined, total mortality, total cardiovascular mortality, stroke mortality, total cardiovascular events (either fatal or nonfatal), or stroke events could constitute the major end points of primary interest. The investigators must determine the end points at the outset of the trial and also decide the amount by which they should change. In the definition of the major objectives their importance must be taken into account, the likelihood that the given changes will be achieved by treatment, and the ease of measurement of the chosen end points.

The importance of the objective

Systolic blood pressure may be easier to measure than diastolic pressure but the investigators may decide that diastolic pressure is more important in determining the future health of the patient. Similarly, a death from myocardial infarction may be more important than the occurrence of an infarct from which the patient recovers. Mortality is usually a more important end point than morbidity and total mortality is a more clear measure of the balance between risk and benefit than mortality from one specific cause. It must be noted that a treatment may reduce one cause of death and increase another.

The likelihood that the objective will be achieved by treatment

An antihypertensive drug may reduce both stroke mortality and total mortality. However, the proportional reduction in deaths will be greater with stroke mortality than total mortality. A trial with stroke mortality as its end point is therefore likely to reach a conclusion more quickly than a trial to detect a reduction in total mortality. The end point of stroke mortality may be preferred in this example for ethical reasons, limiting harm to the control group, and for efficiency. However, total mortality has also to be considered in assessing the benefit : risk ratio and the monitoring of more than one end point is discussed under the decision rules for stopping a trial (chapter 12).

Ease of measurement of the end point

It may be easier to measure systolic blood pressure than diastolic pressure and the former measurement may have greater repeatability. Similarly, the fact of death is easier to determine than a particular cause of death and a non-fatal cardiovascular event may be particularly difficult to ascertain. If the patient dies it has to be decided whether or not sudden death should be regarded as a cardiovascular death, and, if so, how quickly must the patient die to be considered as a sudden death. Also, if the patient survives a myocardial infarction, the diagnostic electrocardiographic or enzyme changes have to be agreed in advance. Everything being equal the end point with greatest repeatability, validity and sensitivity should be chosen (chapter 14).

WHAT CHANGE IN THE END POINT MUST BE DETECTED?

We must distinguish between biological and statistical significance, consider whether the end point is a continuously distributed or qualitative variable, and, if qualitative, whether or not the end point occurs frequently or rarely.

The distinction between biological and statistical significance

The distinction has to be made between what is statistically significant and what is biologically important. It may be observed that an antianxiety drug lowers systolic blood pressure by, say, 2 mmHg and this result, given a large number of patients, could be highly statistically significant. However, the biological importance of the result would be small and the drug would not be used as an antihypertensive agent. In hypertension, a drug is very useful if it lowers systolic blood pressure by more than 10 mmHg. The objective of the trial of a powerful new antihypertensive agent might therefore be to test the hypothesis that the drug lowers systolic blood pressure by 10 mmHg (or conversely, the null hypothesis that the drug does not do so). The greater the biological effect the more likely are the results of the trial to be incorporated into daily clinical practice.

Table 4-1. Average results and standard deviations for the results of a survey on 634 London Civil Servants aged 35-64, 34% of whom were female.

	Mean	Standard Deviation	Biologically Important Change	
Systolic blood pressure (mmHg)	133	20	-10	(-8%)
Diastolic blood pressure (mmHg)	82	13	- 8	(-10%)
Blood haemoglobin (gm/100ml)	14.5	1.1	+ 2	(+14%)
Blood glucose (mmol/l)	5.5	1.4	-1	(-18%)
Serum cholesterol (mmol/l)	6.3	1.0	-0.6	(-10%)
Serum urate (mmol/l)	0.32	0.07	-0.05	(-16%)
Serum creatinine (mmol/l)	93	15	-15	(-16%)

Changes in continuously distributed variables

What sort of changes in continuously distributed variables are of biological importance? Table 4-1 gives some biochemical and other results from a screening of London civil servants [73]. The table also gives some suggestions for changes that could possibly be produced by treatment and be considered biologically important. The suggested changes are of the order of ten to 20 percent or about one standard deviation. For example, a reduction in diastolic blood pressure of 8 mmHg; an increase in haemoglobin of 2 gm/100 ml and a reduction in blood glucose of 1 mmol/liter could all be considered biologically important.

Changes in qualitative end points

When defining a change in a proportion, we may have more difficulty in identifying a biologically important effect as opposed to a lesser effect. For example, if we are considering a reduction in mortality it could be argued that any reduction, however small, is important. On the other hand the cost and adverse side effects of treatment may negate small benefits. In addition, with treatment to prevent an uncommon event, not only should the percentage reduction in the event be considered but also the absolute benefit in numbers/1000/year. It has been suggested that ten young individuals with mild hypertension would have to take antihypertensive medication for 20 years to reach an even chance of avoiding one cardiovascular event. If the probability of an event is low, a treatment used for prevention must be highly effective, whereas a patient with an incurable illness will be interested in a trial treatment that offers a cure only in a small percentage of patients. We shall consider the infrequent and frequent end point in more detail.

The end point occurs infrequently

Cardiovascular events, though not rare in the general population and common in patients with a previous history of cardiovascular disease, may occur very infrequently during a controlled trial. A trial of secondary prevention is intended to prevent a recurrence of a condition and new events may be expected. This is to be contrasted with a primary prevention trial intended to prevent the condition initially where events may be rare. Even with a secondary prevention trial of myocardial infarction, fewer than 15 percent of patients who leave hospital will die over a one-year period. For the purpose of this discussion such events will be considered infrequent.

Table 4-2 illustrates some trials that have suggested a benefit in preventing cardiovascular disease; it also indicates whether or not the authors of the trials have suggested the treatment be adopted; and considers whether the benefits have been accepted and the findings implemented by the medical community at large. The trials included in the table include the Veterans Administration Cooperative Study Group on Antihypertensive agents [52,64] and trials

Table 4-2. The results of several large trials to detect a reduction in cardiovascular disease. The benefits are listed, together with the authors' recommendations and subsequent use of the treatments.

Trial Primary prevention	End point	Reduction	Did authors recommend treatment?

Veterans Administration
  Cooperative study
  group on Antihyperten-
  sive agents [52,64]
	Morbid or mortal
	  events		

Diastolic 115-129 mmHg		93%	Yes[+]
Diastolic 105-114 mmHg		77%	Yes[+]
Diastolic 90-104 mmHg		33%	Yes[+]
Clofibrate [74]	IHD incidence	20%	No
	Non fatal MI	25%	No
	Fatal IHD	(8% increase)	No
	All deaths	(22% increase)	No

Secondary prevention of myocardial infarction

Anturane [65]	Sudden deaths (males)	43%	Yes
Aspirin [75]	Total mortality	25%	No[+]
Anticoagulants [76]	All deaths (males)	20%	Not clear
	All deaths (females)	8%	No
Thrombolytic therapy [77-83]	Total mortality	25%	Yes[+]

IHD = Ischaemic heart disease; MI = Myocardial infarction.
[+] treatment has been widelu used.

designed to test sulphinpyrazone (Anturane), aspirin, clofibrate, and anticoagulants in the primary and secondary prevention of ischaemic heart disease. The 93 percent and 77 percent reductions in morbid or mortal events in the Veterans Administration antihypertensive agents trial have been accepted and acted upon by the medical profession. However, when treatment in this same trial produced only a 33 percent reduction in cardiovascular events, the results were not widely accepted and many physicians do not treat a diastolic pressure of 90-100 mmHg, even though most later trials have also supported treatment in this range.

In secondary prevention of myocardial infarction, an analysis of trial results in 1970 [76] suggested a 20 percent reduction in death in men. Although it can be argued strongly that this is worthwhile, the medical profession as a whole has not considered such benefits warrant the expense and difficulty of long-term treatment. Anticoagulant treatment has therefore fallen from favour and the use of sulphinpyrazone (Anturane) and aspirin was not widely accepted at first, although later trials have strongly supported the use of aspirin [77]. It must be admitted that inconsistencies in the data have not helped. For example, anticoagulants have little effect on the death rate in women and treatment may only be suitable for men. Total deaths were increased by clofibrate (table 4-2) and it would not be acceptable to employ this treatment.

The evidence for benefit from thrombolytic therapy was strong and precise and appeared in the years 1986-88 [77-82]. Nevertheless although the results were widely circulated some regions in the UK were not extensively using this treatment until 1991 [83]. Clinicians tend to ignore small benefits, and when the event rate for a disease is low the patient may also be unwilling to take treatment to reduce a small risk by only, say, 25 percent. The patient will be more interested in therapy that almost guarantees freedom from the disease.

When the end point for the trial is very frequent

If a condition gives rise to a high mortality, as with certain cancers, both the doctor and patient may be interested in a drug that reduces mortality by less than 20 percent. A patient faced with no hope of recovery may be pleased to accept a one in five chance or less of survival.

Other considerations necessary for the definition of major objectives

The major objective must not only include a statement on the nature of the major objectives and the size of the treatment effects that must be detected but also the statistical significance with which the results should be reported, and the confidence intervals for the result (chapter 14). In addition, it is important to limit the number of major objectives in order to avoid multiple statistical testing of several end points. Preferably there should only be a single end point although up to six has been necessary to give a full evaluation of outcome (chapter 14).

SPECIFICATION OF MINOR OBJECTIVES

Minor objectives are subsidiary to the main purpose of the trial and are not used to calculate the number required for the trial. Minor objectives may be specified in order to fully interpret the conclusions originating from the trial. For example, it would make little sense to mount a long-term trial to assess the reduction in stroke incidence (main objective) without assessing the adverse effects of treatment. Minor objectives may be required to assess the trial results or may be only loosely connected with the main purpose of the trial. It is desirable to answer additional questions where possible, provided the smooth running of the trial is not jeopardised. In the previous example a drug treatment may be employed to reduce the incidence of stroke and the opportunity may be taken to examine the effect of the drug on blood lipids. We shall discuss examples of minor objectives and the factors limiting their investigation.

Examples of the minor objectives that can be defined in a long-term trial

Table 4-3 lists some minor objectives that may be formulated during a long-term trial of drug treatment to reduce mortality. The major objective is given first and includes a statement of the size of the effect to be detected, the level of significance to be achieved and, in the event of the effect not being demonstrated, the confidence with which it is excluded (power).

The essential minor objectives in this type of trial will include an assessment of both total mortality and any serious but non fatal adverse effects of treatment. The objectives can be illustrated by considering again the European Working Party trial of Hypertension in the Elderly (EWPHE) [60]. The major objective was defined as a reduction in stroke events (mortal plus morbid events) of 50 percent, with a level of significance of 5 percent and a power of 90 percent. Certain minor objectives listed in table 4-3 have been studied and resulted in a description of the natural history of untreated hypertension in the elderly and an examination of the changes in cardiac and renal function with

Table 4-3. Possible objectives that may be defined during a trial of drug treatment versus placebo in reducing mortality from a chronic disease.

Major objective

1. Does the drug treatment reduce mortality from disease V by X% (with a significance of Y and a power of Z)?
2. The effect of drug treatment on total mortality will also be examined but it is not expected that statistical significance will be achieved. Any serious adverse effects of drug treatment will also be determined.

Minor objectives

1. To determine the natural history of the disease while on placebo.
2. To identify any biochemical effects of drug treatment.
3. To identify any symptomatic side effects of drug treatment.
4. To determine the interrelationships between the condition under treatment and other diseases.
5. Objectives unrelated or only indirectly related to the condition being treated (e.g. factors influencing compliance with therapy).

---

increasing age in these subjects [84]. In addition, the biochemical changes with active treatment were estimated and a reduced glucose tolerance in the group treated with a diuretic reported [85]. Other minor objectives listed in table 4-3 include the association between the condition under treatment and other diseases, the side effects of treatment, and items of general interest such as the factors influencing compliance with treatment or default from follow-up.

Factors limiting the investigation of minor objectives

The greater the number of objectives, the greater will be the complexity of conducting the trial. A trial with many minor objectives may require repeated biochemical and other investigations and impose a greater burden on the investigators and the subjects so that both will be required to make more time available for the trial.
Additional expenses will also be incurred and clerical tasks increased. However, without answering several questions, it may be difficult to assess the results of a trial in terms of benefit and risk (chapter 17).

CONCLUSIONS

It is very important to specify the major objectives for a trial. They must include the most important end point to be measured, the size of any treatment effect that has to be determined, the significance with which the effect must be observed, and the power of the trial in the event of a nonsignificant result being reported. Armed with this information and an estimate of the variance of the end point under consideration, the numbers required for the trial can be calculated (chapter 6).

Minor objectives may have to be specified in order to fully appreciate the outcome of the trial. Some will be necessary for this purpose and others will be incidental to the main objective of the trial. How many minor objectives are defined will depend on the resources available, but the collection of a large amount of information during the course of a trial may hinder its successful completion.

5.   DIFFERENT TRIAL DESIGNS

In the standard trial design a subject or patient is considered suitable for entry to the trial, is then randomised to a treatment group and each group receives a specified treatment. A fixed number of persons enter the trial and are followed for a predetermined interval of time; treatment is stopped at the end of this period. Other trial designs have to be considered. The subject may be asked to take more than one treatment consecutively (cross-over trial), more than one treatment simultaneously (trial to detect an interaction between treatments), or more than one treatment consecutively and concurrently (cross-over trial to detect an interaction). The remaining design variations do not involve a fixed number of persons being entered. The standard trial, cross-over trial, trial to detect an interaction between treatments, and cross-over trial to detect an interaction are illustrated in figure 5-1 for four treatments A, B, C, and D.

The standard design (number 1) allows four treatments to be compared simultaneously and has been called a parallel-groups trial. If one treatment is a placebo then the effect of the other three treatments can be estimated in the full knowledge of any placebo effect or change in baseline measurements. In the cross-over trial (number 2) each subject receives the four treatments and the order of treatment is randomised. In this example four different orders are specified. Trial design number 3 allows the effect of one treatment to be assessed in the presence of a second treatment (that is, the presence of an interaction between treatments may be determined). The individual drug effects are also determined as a fourth group receives neither drug (but usually receives a placebo). In design 4 an interaction between treatments can be determined within-subject in a cross-over trial.

58

1. Standard trial design

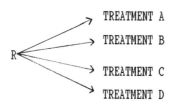

TREATMENT A

TREATMENT B

R

TREATMENT C

TREATMENT D

2. Cross-over trial

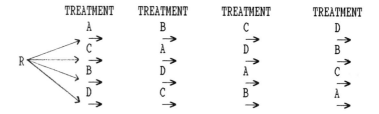

3. Trial to detect an interaction

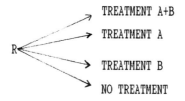

TREATMENT A+B

TREATMENT A

R

TREATMENT B

NO TREATMENT

4. Cross-over trial to detect an interaction

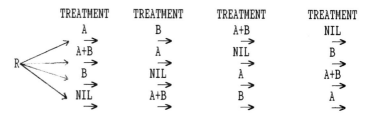

Figure 5-1. Two designs to assess the effects of four treatments A,B,C, and D two designs to test for an interaction between two treatments A and B. R = point of randomisation.
1. Standard trial design. Four groups studied in parallel to assess four treatments.
2. Cross-over design to assess the effects of our treatments within patients.
3. Parallel groups study to detect an interaction between two treatments.
4. Within-patient cross-over trial to detect an interaction.

THE USE OF THE STANDARD TRIAL DESIGN

The standard design is the one most commonly used and has the virtue of simplicity in that a single treatment or combination of treatments is given to each group and a fixed number of patients is involved. The other frequently employed design, the cross-over trial, is inappropriate when a treatment is curative, when the duration of treatment has to be long, when the effects of treatment persist for some time after stopping treatment (a carry-over effect), or when a large number of treatments have to be compared. In any of these circumstances the standard design has to be employed. This standard design is also more appropriate when a large number of subjects are available for the trial.

When should standard parallel groups design be employed?

When treatment is curative

If the trial is to test a curative treatment for an illness, the cross-over design cannot be employed. There is no point in a cured patient continuing with further treatments.

When the duration of treatment has to be long

If the effect of a drug has to be determined after, say five years, a cross-over trial may take too long since the duration of the trial is ten years with two treatments and 20 years with four.

When the effect of one treatment is different when it follows another treatment

If the effect of one drug persists for a long time, the carry-over effect of this treatment into the next treatment period may interfere with the effect of any further treatment and the standard trial design is to be preferred.
When comparing two treatments in a cross-over trial the difference between the treatment

effects must be independent of the order of administration. Hills and Armitage have concluded that if previous experience with the treatments has not proven that this is true, then a parallel group study should be carried out [86]. A different result in one period (in statistical terms, an interaction between treatment and period) may not only occur with a carry-over effect. For example, placebo treatment may possibly be more effective when given first to lower blood pressure or when given last to relieve a painful condition that is improving with time.

When a large number of treatments are to be compared

With a large number of treatments the trial would be too long and complex if each subject had to take every treatment. The standard trial allows several treatments to be compared.

When the number of subjects available for the trial is unlimited

In certain circumstances (see below) the cross-over trial will be more efficient, requiring fewer subjects. This may be true if the variation of between-subject measurements is much greater than the variation of within-subject measurements. However, when large numbers of subjects are available for the trial we can assume that enough subjects will be recruited in a standard design to detect an important difference between the treatments. Assuming costs are not the limiting factor, we can opt for the less efficient design but one that is quicker and simpler to execute without the difficulty of possible carry-over effects. As the trial duration is shorter, fewer patients will drop out in proportional terms.

THE USE OF THE CROSS-OVER TRIAL DESIGN

The cross-over design can be considered when the condition being investigated is constant and only temporarily affected by treatment. For example, a patient with a high blood pressure or blood sugar may receive drug treatment that has a short term effect on his condition and may then take a succession of treatments that do not affect the result of later treatments.

Cross-over trials may be recommended when any carry-over effect is short, when the prolongation of the trial neither greatly increases dropout rates nor alters the relative effects of the treatments being compared, when the within-subject variation is less than between-subject variation, and when any order effect can be balanced out.

Warning. It is difficult to prove that the difference between treatment effects is independent of the period of treatment and therefore the Food and Drug Administration in the United States has concluded that the cross-over trial is not the design of choice where unequivocal evidence of treatment effect is required. Hills and Armitage also concluded, *"If the number of patients is limited and a cross-over design is chosen, then the internal evidence that the basic assumptions of the cross-over are fulfilled must be presented and if necessary the conclusions should be based on the first period only"* [86].

The investigator may, however, be certain from previous studies that the difference between treatments is independent of period and he can then proceed with a cross-over design. It must be remembered that a difference between treatment effects may be due to a carry-over effect of one treatment into the next period or to an influence of the time of assessment from the beginning of the trial, a so-called order effect.

The cross-over trial should be considered when:

Any carry-over effect is of short duration

In the treatment of hypertension with antihypertensive drugs the carry-over effect in lowering blood pressure is usually short, and a brief interval between treatments will ensure that one treatment does not influence the result of the next. The interval may last from two to

four weeks and has been called a wash-out period.
However, certain antihypertensive drugs may have
longer effects on measurements other than blood
pressure (for example, diuretics may reduce serum
potassium for three months). With such long
effects the cross-over trial design may not be
appropriate, but under certain stringent
conditions the trial may still be analysed with
the carry-over effect balanced out and even
estimated (chapter 13).

Extending the treatment period does not alter the difference between the treatment effects

As discussed earlier it is possible that the
difference between the two treatments may differ
at the start of a trial from later periods in the
trial. Meier and Free [87] have therefore argued
that *"each patient as his own control"* is not entitled
to the status of dogma. They reviewed the results
of cross-over studies on the use of analgesics in
postoperative pain. In this situation the pain is
lessening with time and the standard design has
the advantage of simplicity. However, the
between-patient differences are considerable,
supporting the use of a cross-over trial, and it
is possible to allow or adjust for the order
effect in the design or analysis. Although the
treatment effects may diminish through time the
difference between various treatments may be more
consistent.

A baseline measurement cannot be made

In a parallel-groups trial, precision may be
increased when the within-subject variance is
lower than that of the between-subject variance
and baseline measurements are employed to provide
within-subject data. However within-subject
changes on a single treatment may be related to
those on a second treatment and the concept of
'the patient acting as his own control' is more
applicable for cross-over trials than parallel
groups trials. The cross-over design may be
expected to be more sensitive to differences
between treatments. Moreover, if baseline
measurements cannot be made, the parallel-group

trial will require many more patients than a cross-over trial. One example would be provided by patients with severe diabetes mellitus who require insulin treatment every day. If two new insulins are to be compared for their effect on blood sugar, the baseline blood sugar would be unsatisfactory as the current treatment cannot be stopped and the starting blood sugar represents the efficiency of the original treatment. However, if two diets were to be compared *in addition to* the original treatment then the starting sugar would be a satisfactory baseline measurement.

A *cross-over trial will not result in a large increase in dropouts*

The longer an individual takes to complete a trial the more likely that the person will default (dropout). This problem is discussed in chapter 11. The dropout rate will increase with time owing to the subjects moving address, having an intercurrent illness, changing their occupation, and taking holidays. All these possibilities increase with the duration of the trial and, in addition, dropout may occur owing to the subjects becoming intolerant of the number of visits, repeated investigations, or one of the treatments employed. If an adverse effect of a treatment is experienced in the first or second treatment period the patient may be unable or unwilling to take further treatments.

When the treatment period is three months, it will reduce the dropout rate per treatment to give four times a given number of patients one of four treatments for three months than to give the patients all four treatments over a one year period. However, it may be difficult to recruit four times as many patients for the standard trial design and the costs of recruitment and initial investigation will be increased.

An *order effect is absent or can be balanced out*

The order effect is the change in a measurement according to the period of estimation after allowance has been made for the effect of treatment. In a trial of antihypertensive drugs blood pressure can become progressively lower as the trial proceeds. The exact mechanisms producing this fall in pressure have not been determined. Initially, pressure falls due to familiarisation with the technique of measurement, the observer, and the surroundings. This effect can be reduced by a prolonged run-in period prior to randomisation. Additional reasons have been suggested for the fall in pressure: an effect of any placebo tablets that are given; a phenomenon whereby an initial lowering of pressure makes blood pressure control easier thereafter; and the removal from the trial of persons whose blood pressure rises leaving a higher proportion of those in whom pressure falls. Whatever the cause of a trend with time the subjects must be randomised to receive the treatments with equal frequency at different times. In figure 5-1 and trial design 2, if equal numbers of patients are randomised to the four sequences, then drug A will be as often given first as second, third, or fourth. Similarly with drugs B, C, and D, the order effect for these treatments is said to be balanced out. The order effect can be estimated by comparing the average results for each period, every interval including an equal number of measurements on the four treatments.

The order effect may be important in trials other than those of antihypertensive drugs or analgesics. In the treatment of diseases with a fluctuating course the trial may be commenced when the condition is at its worst and a subsequent improvement is expected as part of the natural course of events.

THE STUDY OF TWO OR MORE TREATMENTS SIMULTANEOUSLY: FACTORIAL DESIGN OR TRIALS TO DETECT AN INTERACTION BETWEEN TREATMENTS

Traditionally, in an investigation the experimenter isolates a number of factors and studies the result of altering one factor while holding the others constant. Fisher considered this doctrine to *"be more nearly related to expositions of elementary physical theory than to laboratory practice in any branch of research"* [88] and we shall consider the advantages of more complex experiments where two factors (treatments) are given together. The simultaneous examination of more than one treatment allows any interaction between the treatments to be determined. If an interaction is not present the experiment allows an extra estimate of the two treatment effects.

The detection of an interaction between the treatments

An interaction is said to be present when the effect of one factor is different in the presence of another factor. Let us consider a trial where the subjects receive either drug A, drug B, A plus B, or placebo (figure 5-1, number 3). This design is know as a factorial experiment and yields two estimates of the effect of two factors, drugs A and B, one estimate of the drug's effect when given alone and one of its effect when given in combination. Let us suppose that the drugs lower serum uric acid, the mean uric acid after drug A being U_A, after drug B, U_B, after both drugs combined, U_{A+B}, and when on placebo, U_O. Figure 5-2 provides fictional data for a factorial design. The upper lines connect the results when treatment A was not taken and the lower lines when it was. The left-hand results were obtained when drug B was not given and the right-hand results when B was taken. The figure illustrates three sets of results: the upper panel when no interaction is present; the middle panel when a negative interaction occurs, and the lower panel when a positive interaction is demonstrated. When there is no interaction the effect of A is the same irrespective of the presence of B and vice versa and the distance between the two lines gives the effect of drug A, equal to $(U_A - U_O)$ or $(U_{A+B} - U_B)$. The effect of drug B is $(U_B - U_O)$ or $(U_{A+B} - U_A)$. When an interaction is not present the drug effects are said to be additive. In the presence of an interaction the difference estimates of the drug effect are not equal.

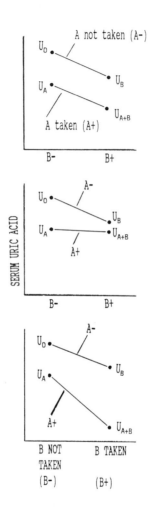

Figure 5-2.   The effect of two drugs A and B on serum uric acid. The left-hand results are when B was not taken and the right-hand results when B was taken. The upper lines represent the situation when A was not taken and the lower lines when A was consumed. The upper graph illustrates the situation when no interaction between the treatments was present, the middle when a negative interaction was observed, and the lower graph when a positive interaction was present (see text).

The middle graph in figure 5-2 illustrates the result when the two drugs in combination have less than the expected additive effect. This is known as a negative interaction and has been loosely referred to as antagonism between the drug effects. The lower graph represents a positive interaction when the effect of the drugs in combination is greater than expected. The term *synergism* has unfortunately been used both for an additive effect (with no interaction) and a multiplicative effect (with a positive interaction). Few trials to detect interactions have been performed; early examples are given by Wilson and his colleagues and Aenishänslin and co-workers [89,90].

More than one estimate of the treatment effects

The standard design would consist of a group on drug A, a group on drug B, and a group on no treatment. This design would give one estimate of the effect of drug A and one estimate for drug B. Overall, this design yields two estimates for three treatment groups, whereas the factorial design gives two estimates of each drug effect, four estimates for four treatment groups. The factorial design is more efficient and does not lose precision [88]. However, if the effect of drug A is not the same in the presence of drug B (and vice versa) then the factorial design gives only one estimate of the effect of each drug but it does detect the interaction. As discussed previously the estimate of the effect of drug A is given by the two comparisons $(U_A - U_o)$ and $(U_{A+B} - U_B)$ and similarly the effect of drug B is given by $(U_B - U_o)$ and $(U_{A+B} - U_A)$. The overall estimate of the effect of A is given by

$$\frac{(U_A - U_o) + (U_{A+B} - U_B)}{2}$$

and the effect of B by

$$\frac{(U_B - U_o) + (U_{A+B} - U_A)}{2}$$

This trial design should be more widely used, especially when two treatments are thought to have moderate, but additive, effects (for example, a reduction in mortality of 10 percent for each drug). A trial including both A and B simultaneously may be able to detect a 20 percent reduction in mortality while a standard design of treatment with A alone, B alone, and placebo may fail to detect a reduction of 10 percent.

CROSS-OVER TRIALS TO DETECT AN INTERACTION

Figure 5-1, design number 4, gives an example of this trial design which is a combination of a within-patient cross-over design and a design to detect an interaction. In such a trial the individual patient has to receive, for example, four different treatments in a certain order. The order can be randomised so that for each set of n patients the order effect and carry effects are cancelled out (see below).

Such a design may consist of n treatments arranged in one or two n X n squares using Latin letters to designate the treatment. Such a design is know as a Latin Square and is appropriate for the cross-over design to detect an interaction between treatments and for other trial designs.

The design has been utilized successfully for antihypertensive drugs [91,92,93] and antianginal agents [94]. Owing to the lower within-subject variance of measurements of blood pressure and frequency of angina, these trials gave estimates of drug effects with very few patients. However, the trials did not detect interactions between treatments either because they were absent or because the trials were too small to detect these effects.

THE LATIN SQUARE OR RANDOMISATION SUBJECT TO DOUBLE RESTRICTION (ROW AND COLUMNS)

Figure 5-3 gives two 3 X 3 Latin squares. the first square gives the order of administration of three treatments A, B, and C to three patients designated 1 to 3, and the second square, for a

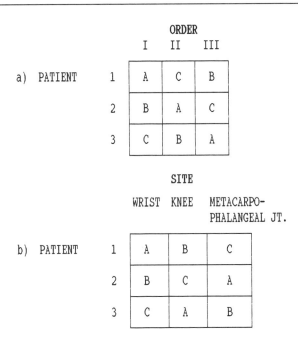

ORDER

		I	II	III
a) PATIENT	1	A	C	B
	2	B	A	C
	3	C	B	A

SITE

		WRIST	KNEE	METACARPO-PHALANGEAL JT.
b) PATIENT	1	A	B	C
	2	B	C	A
	3	C	A	B

Figure 5-3. Two 3 X 3 Latin squares to allocate each treatment to every patient and to ensure that each treatment is used once at each order or site of administration.

different trial, gives sites for the application of three treatments. With the first square, for every three patients each treatment is given first, second, and third once only; the design is said to balance out any order effect. The design also ensures that each patient receives all three treatments. The second square could be useful in a trial of three intra-articular injections in patients with severe generalised rheumatoid arthritis. The most severely affected wrist, knee, and metacarpo-phalangeal joint could be selected, and in the first patient, injection A made into the wrist, B into the knee, and C into the metacarpo-phalangeal joint. For every set of three patients each type of joint will receive all three treatments. The design could be said to be balanced for the joint treated.

Randomisation of patients in a Latin square design

Let us suppose that 18 male patients are to be randomised to a Latin square design with three treatments and three orders of treatment. Three squares identical to a) in figure 5-3 could be taken and three equal to b). These squares can be pooled to an 18 X 3 table, and the rows numbered one to 18. The patient would be randomised to these rows. For example, the first patient may be randomised to row three, the second to row 17, and so on. The randomisation can be easily read from a random number table (chapter 7) by noting a sequence of numbers less than 19. Randomisation will ensure that the investigator cannot predict the order of treatment for the patients entering the trial. However, he could (but only if he wished) predict the eighteenth order.

The treatment and order effects

The differences between treatments are calculated by comparing the average results for the patients when they are taking the particular treatments. Similarly the average result for the orders is derived from an equal number of the different treatments.

The carry-over effects

The effects of a treatment may continue into the next period; this is known as a carry-over effect (see above). Latin square designs can be employed that balance out residual or carry-out effects [95-96]. If the number of treatments is even, one square can be designed to achieve this effect, and if the number of treatments is odd two squares are required. Only certain Latin squares have these characteristics. Figure 5-3 gives the two squares required for three treatments. When the two squares are combined each treatment follows every other treatment twice. Figure 5-4 gives the one square required for four treatments and the two squares necessary with five treatments.

ORDER

		I	II	III	IV
PATIENT	1	A	B	D	C
	2	B	C	A	D
	3	C	D	B	A
	4	D	A	C	B

ORDER

		I	II	III	IV	V
PATIENT	1	A	B	D	E	C
	2	B	C	E	A	D
	3	C	D	A	B	E
	4	D	E	B	C	A
	5	E	A	C	D	B

ORDER

		I	II	III	IV	V
PATIENT	6	A	C	B	E	D
	7	B	D	C	A	E
	8	C	E	D	B	A
	9	D	A	E	C	B
	10	E	B	A	D	C

Figure 5-4. Latin square designs to balance out order effects and carry-over (residual) effects. In order to achieve balance, one square is required for four patients and four treatments, and two squares for ten patients and five treatments.

When single squares balanced for carry-over effects are duplicated or a combination of Latin squares is employed that is balanced the residual effects can then be estimated. The methods for calculating the residual and other effects havebeen clearly described and examples of balanced square provided for more than five treatments [96].

THE GRAECO-LATIN SQUARE

The Graeco-Latin square employs both Latin and Greek letters and allows three different sources of variation to be equalised. For example, a trial may be designed for three drug treatments, three patients, three orders, and three methods of administration (oral, intramuscular, and intravenous). Figure 5-5 gives an example of such a trial. In this Graeco-Latin square each drug treatment A, B, and C is given once orally (α), once intramuscularly (β), and once intravenously (γ). More complex Graeco-Latin squares together with their methods of analysis have been described by Cochran and Cox [96].

		ORDER		
		1	2	3
PATIENT	1	Aα	Bβ	Cγ
	2	Bγ	Cα	Aβ
	3	Cβ	Aγ	Bα

Figure 5-5. A Graeco-Latin square for three patients; three drug treatments A, B, and C; and three orders of administration, α (orally), β (intramuscularly), and γ (intravenously).

THE SEQUENTIAL TRIAL

Armitage defined a sequential trial as a trial where *"Its conduct at any stage depends on the results so far obtained"* [97]. Usually a sequential trial compares two treatments, and the results during the course of the trial determine the number of observations made. In most trials of any design patients are started on treatment serially and not simultaneously; therefore it is possible to assess the response to treatment as it becomes available in sequential order.

A section of statistical theory termed *sequential analysis* derives largely from the work of Wald [98] who allowed for repeated significance testing and derived boundaries describing three possible outcomes. Figure 5-6 gives an example of such boundaries where a comparison is made between drug T and drug A. The upper boundary is a boundary that must be reached to demonstrate a statistical preference for T and the lower boundary must be reached to demonstrate a preference for A. If the two boundaries forming a V shape at the right centre of the figure are reached, then the investigator knows that a preference for one drug over the other is not likely to be demonstrated within the predetermined conditions of the trial. A design with this central limiting boundary is known as a closed design. Figure 5-6 illustrates the result of a trial by Robertson and Armitage [99] where two hypotensive drugs used during operations were compared, T being phenactropinum chloride (Trophenium) and A being trimetaphan (Arfonad). The comparison was made between pairs of subjects, and for each pair the time taken for the systolic blood pressure to rise to 100 mmHg after the use of the drug was measured. The results are plotted and do not reveal a preference for either drug.

Figure 5-7 illustrates the results when two cough suppressants, heroin and pholcodine, were compared with each other and placebo. Comparisons were made within subject [100] and after six days the patients had tried all three treatments and ranked them in order of preference. The trial was designed to detect a significant difference between pairs of the treatments at the five percent level of significance with a power of 95 percent when 85 percent of preferences are in favour of one drug. Both heroin and pholcodine were preferred to placebo but no distinction could be made between the two active drugs.

The decisions that have to be made to employ a sequential trial design

As in a standard trial the levels of type I (α) and type II (β) error have to be decided (chapter

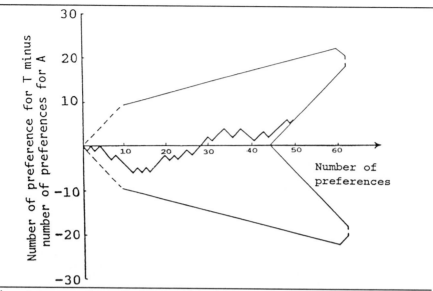

Figure 5-6. The result of a sequential trial to compare two hypotensive drugs employed in anaesthesia, phenactropinum (Trophenium or T), and trimetaphan (Arfonad or A). Reproduced with permission from Robertson and Armitage, Anaesthesia 1959;14:53.

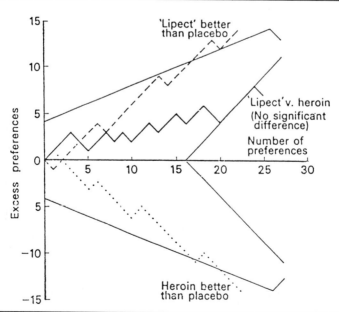

Figure 5-7. A trial to compare pholcodine (Lipect), heroin, and placebo as cough suppressants. Reproduced with permission from Snell and Armitage, Lancet 1957;1:860-862.
---- Comparison of Lipect with placebo.   .... Comparison of heroin with placebo.
___ Comparison of Lipect with heroin.

6). In addition the trial may be of an open or closed plan. In other words, the investigator has to decide whether he is prepared to allow the sample size to increase indefinitely, or whether he will restrict the trial so that if a specified difference between treatments is not apparent by a certain stage, then the trial is stopped. Figures 5-6 and 5-7 illustrate closed plans. Thereader is referred to Armitage [97] for further details. Armitage recommends a closed plan for medical trials as an unexpectedly long series of observations may be a considerable disadvantage. The closed plan reduces the maximum possible sample size.

Advantages of sequential trials

Sequential trials have the ethical advantage of terminating quickly when one drug is an important new advance. These trials may also prove economical and useful as a pilot to determine the variance of the measurements.

Ethical advantages

The investigator may wish to follow the results of the trial closely and continuously and bring the trial to an end immediately when any statistically significant difference is observed: for example, a new treatment for cancer where the treatment is widely available. The investigator will wish to reach a conclusion in the shortest period of time in order to ensure that the treatment, if successful, is generally applied.

Economy

If the trial is brought to a speedy conclusion, then the financial cost of the experiment will be reduced. This will be true only when one treatment is much worse than another.

Use in pilot studies

Anscombe [101] has suggested that if the numbers required for a nonsequential trial cannot be calculated because the variance of the measurements is unknown, then a pilot study may be conducted in a sequential manner until an estimate of variance has been determined with a given precision. The second stage can then be a standard trial or a further sequential trial.

Disadvantages of the sequential trial

The sequential trial is not suitable for long-term studies or when secondary objectives are important, and the method does not guarantee economy.

Difficulty of use in long-term studies

The objective of many sequential trials is to bring the study to an end before many treatments have been started. If the period of the observation is long in comparison to the time taken to enter patients in the trial, then there is little scope for limiting the number of patients who do enter the trial. A sequential trial is most appropriate when the response is obvious soon after treatment is started. Sequential trials are therefore more suitable for the treatment of acute leukaemia than, say, Hodgkin's disease, which has a more prolonged and fluctuating course.

Loss of additional information

In the standard trial the larger number of patients may allow end points to be reported with smaller confidence limits and also may allow more observations on related aspects such as side effects or the more severe adverse effects of treatment. A sequential trial is therefore not appropriate when important secondary objectives have been defined.

Economy may not be achieved

We must remember that the sequential plan is more economical on average than a nonsequential trial. However, in exceptional circumstances the sequential trial may require more patients and observations than a standard trial. If one treatment is only moderately worse than the other the standard procedure and analysis may be more likely to give a statistically significant result, one final test being performed rather than a repeated sequence of tests (chapter 6).

Organisational problems

A sequential trial, by definition, will last for an unknown duration. It is therefore more difficult to estimate the total cost of the trial or to know how long to employ staff working on the trial.

Concealment of the results during the course of the trial

The statistician in charge of the sequential analysis will naturally plot the results of the trial graphically. If this clear graphical representation is seen by the clinicians involved in the trial, they will naturally have an idea of the likely result of the trial. As a boundary is approached they will imagine that one drug has superiority over another. This conviction will bias their attitude to the trial and may lead to a demand that the trial be stopped. These problems may be overcome by making sure that only a central monitoring committee has access to the results during the course of the trial.

Other possible disadvantages

Cochran [102] stated, "In the sequential trial, at the beginning, the doctor is forced to make some decisions about the desired sensitivity of the trial which he can dodge in a fixed-size trial." However, if the researcher is going to estimate the number of subjects he will require for his fixed size trial, then he must make the decisions in the same way as he would for a sequential trial. It is hoped that investigators will not set out on a trial without prior consideration of whether they will recruit sufficient patients for their purpose.

The investigator embarking on a sequential trial will be attracted by the economy in both the subjects involved and the observations required. In addition to the sequential strategy he may also include a low figure for the power term. This will reduce confidence when the boundary is crossed for a nonsignificant difference between the treatments. Although the sequential design will not differ in this respect from a small nonsequential trial with limited power, there will be a tendency for the trial employing a sequential design to include a lower specification for power. The tendency to use a low-power term in a sequential trial must be avoided. Armitage [97] pointed out that a negative trial that is obviously low in power may inhibit further work, *"either because other investigators attach more importance to the first negative results than they deserve ... or because they have less enthusiasm for repeating previous work than for breaking entirely new ground."* Lastly, reporting the results of a sequential trial should not consist solely of a graphical representation. A summary of the data as a whole must still be presented with means and standard errors, comparisons between relevant subgroups of treatments, and confidence limits.

Conclusions on sequential trial

Despite the difficulties in the design and execution of a sequential trial, the design should be utilized more extensively. If a sequential trial is started and if one treatment proves greatly superior, much may be gained. When the trial fails to reveal a significant difference between the treatments the investigator can calculate the confidence limits of any possible beneficial effect. Armed with this new knowledge he may or may not proceed to a fixed sample trial of known duration.

PLAY THE WINNER

The play-the-winner trial has been proposed to limit the number of patients who receive an inferior drug during a clinical trial [103]. A simple example would be to keep using one drug until it first fails and then switch to the second drug until it fails and so on.

Problems with a play-the-winner trial

In 1979 this method had not yet been tried in a clinical trial and Meier claimed that *"This is testimony to the triumph of good sense over irrelevant theory"* [104]. There were two main reasons why the method had not been employed.

The population of interest is not limited to the group in the particular trial

Meier considered that if the numbers of patients receiving the inferior treatment are minimised in the group involved in the trial, it may take longer to get enough of them to come to a reasonably sure conclusion. During this prolonged interval, patients in other centres may receive the inferior treatment and suffer the consequences that the trial seeks to avoid [104].

The investigator may select patients who would respond to a particular drug

Chalmers [105] suggested that playing the winner will mean that the investigator having successes with drug A will expect the next patient to receive drug A even when he does not know what drug A is. Allocation to treatment is, therefore, no longer blind and Chalmers stated, *"It is very easy for a self-fulfilling mechanism to get started in which the winner is ahead and only the winning patients are more and more accepted for the study, thus ensuring that the leader is confirmed as the winner."* However, this argument assumes that the investigator can select winning patients for the trial (that is, patients who would respond well to a particular drug). Presumably these patients could have a mild form of disease or other so-called winning characteristic.

Present situation

Despite these reservations two play-the-winner trials have now been reported. The first compared enoxaparin with dextran 70 for thrombosis prophylaxis in gastrointestinal surgery [106] and the second compared enoxaparin with low-dose heparin [107]. If deep venous thrombosis, pulmonary embolism, excessive bleeding or a severe adverse drug effect did not occur within 7 days the patient was classified as a 'winner'. In the first trial of 327 patients, 17% of the enoxaparin group were 'losers' and 25% in the dextran group. The cumulative number of winners determined the allocation of the next treatment and 200 received enoxaparin and 127 dextran 70. In the second trial the treatments were equivalent with 92 allocated to enoxaparin and 91 to heparin. The authors discuss the problem of allocating treatment while some patients have not reached an end point; the problem of having to re-classify patients who changed from winners to losers and vice versa after 30 days; and the problem that traditional statistical methods cannot be applied to these non-randomised data [107]. They support play-the-winner trials when the proportion of losers is high as less patients will be required than in a randomised trial.

Conclusions on play-the-winner trials

A play-the-winner trial is one method of adaptive allocation of patients where the results during the trial determine the treatment to be given to the next patient to enter the trial. A more optimal strategy would be to continuously calculate the probability of success with the two treatments and to allocate future treatment on this basis - e.g., the so-called two-armed-bandit problem, where the arms of a slot machine are the treatments and inserting a coin is treating a patient [108,109]. All these procedures assume that all patients are the same; since this is not true it is safer to randomise and have comparable groups from which conclusions can be drawn. Despite this reservation there could be ethical advantages for the investigators involved in this form of trial.

N-OF-1 RANDOMISED CONTROLLED TRIALS

n-of-1 trials are conducted within a <u>single</u> patient. The patient is randomised to different treatments in a cross-over design and the result is <u>only applicable to that individual</u>. Guyatt and his colleagues (110) have suggested that these trials be performed when:

i)   the condition is so rare that multicentre collaborative trials are not feasible.
ii) when the result of a randomised trial may not be appropriate to a patient because they did not meet eligibility criteria or there exists other evidence that the trial result may not provide the answer in a certain individual.

n-of-1 trials may be performed when the best treatment for the individual is not known, the disorder is chronic and stable, and the treatments to be compared have a rapid onset without prolonged carry over effects (111); similar criteria for the performance of any cross-over trial. On average the trials compared three pairs of treatments.

In a review of 73 n-of-1 trials Guyatt *et al* concluded that 66% provided a definite clinical answer, and original treatment plans were changed in 39% (110). Statistical criteria were met only in 27% of trials.

With adequate resources, a case can be made for n-of-1 trials in an individual. They may or may not prove better than the clinician employing his 'best bet' as first line treatment and proceeding to his second and third preferences if necessary. However, more controversial, is the accumulation of n-of-1 trials to give an idea of the best treatment for the generality of patients. Jaeschke *et al* (112) compared amitriptyline with placebo in 23 n-of-1 trials in fibromyalgia. In 7 trials a clinically definite result could not be reached and in the remaining 16 trials, amitriptyline did not benefit the patient in 10 trials. Thus the trial did not indicate the place of amitriptyline in the treatment of fibromyalgia. Nevertheless *if* there is a small subset of patients who benefit, the n-of-1 trial may be useful. Similarly, March and colleagues compared paracetamol with a non-steroidal anti-inflammatory drug, diclofenac, in the treatment of osteoarthritis (113). Twenty five n-of-1 trials led to a clear difference in favour of diclofenac in 7 patients and no preference for paracetamol. In one analysis the results were graphically displayed for all patients, revealing a 4:1 preference for diclofenac, however 9 patients had adequate symptom control with the safer drug, paracetamol. In a commentary on this trial Campbell (114) considered that n-of-1 trials are most appropriate *"when the treatments have already been well investigated in clinical trials. They are not a substitute for a properly controlled phase III trial to decide the efficacy of a new treatment."* For further discussion please see references 115-117. A series of n-of-1 trials is a seductive proposition but as Lewis has pointed out *"it is surely more efficient to plan a co-ordinated statistical design for the whole series.... the cross-over trial, albeit with a large number of potential treatment periods per patient"* (116).

CONCLUSIONS ON DIFFERENT TRIAL DESIGNS

In this chapter the advantages and disadvantages of the standard parallel groups and cross-over trial designs have been discussed. Cross-over trials tend to be more efficient but the results can be difficult to interpret in the presence of persistent carry-over effects. Factorial designs both within and between subject have been discussed. These designs are very efficient and allow interactions between treatments to be detected. Latin square and Graeco-Latin square designs allow cross-over trials to be performed that are balanced for order and carry-over effects. Lastly, the advantages and disadvantages of sequential trial designs have been indicated and the concept of play-the-winner trials introduced.

When in doubt, it is safest to employ the standard trial design of one treatment for each person and a fixed number of subjects. However, the numbers required may be greatly reduced by a cross-over or sequential design. When two active treatments are to be tested, a factorial design may well prove the most economical. Factorial designs are now proving successful in very large simple trials of treatment, such as the ISIS trials (77).

6. HOW MANY SUBJECTS ARE REQUIRED FOR A TRIAL?

A trial must be large enough to detect clinically important differences between treatments. The number of subjects required depends on the objectives defined for the trial; the level of significance that must be achieved if the objective is reached; the confidence with which a result will be reported if the objective is not reached; and the variability of the end-point measurements.

THE SIZE OF THE OBJECTIVES

The major objectives will be the differences to be detected between the intervention and control group. These objectives are discussed in chapter 4 and may be in absolute units (for example, a reduction in diastolic blood pressure of 8 mmHg) or a proportional reduction in an event (for instance, a 50 percent reduction in stroke mortality). The difference should be both biologically important and capable of achievement. If the average antihypertensive drug can lower diastolic blood pressure by 8 mmHg when compared with placebo and preliminary studies on a new drug reveal a similar effect, it would be sensible to set a reduction of 8 mmHg as the objective in a trial of this new drug and unreasonable to expect a much greater reduction. Even a reduction of 4 mmHg will be biologically important and may constitute an acceptable objective. Similarly a reduction in stroke mortality of 50 percent is high but has been exceeded in a trial of antihypertensive treatment [52] and therefore may be a reasonable objective. Usually the investigator will be interested not only in whether the new treatment is better than the control treatment by a given amount, but also in whether the new and control treatments have a similar effect, or indeed if the new treatment is worse than the control treatment by the defined amount.

The traditional objective: to determine if the new treatment is better, the same, or worse than control treatment

The traditional objective determines: (1) whether the effect (A) of a new treatment, is better than the effect (B) of the control treatment by a given amount; (2) whether the new effect A is comparable with B within certain limits; and (3) whether the unexpected occurs and B is better than A by the given amount. For example: Let the effect of a new drug in lowering diastolic blood pressure be a fall in pressure of A mmHg. Let the effect of control treatment be a fall in pressure of B mmHg. The investigator is interested in whether:

1. $A - B \geq 8$ mmHg

2. $A - B$ lies between 7 and -7 mmHg

3. $A - B \leq -8$ mmHg

These are the traditional aims of a trial but Schwartz, Flamant, and Lellouch [118] have pointed out that there are circumstances where the investigator is only interested in 1. (he only wishes to determine when the new treatment is to be preferred and is not interested if the new treatment is the same as or worse than the old). In certain circumstances the investigator may be interested in 1. and 2. but not 3. (that is, he is not interested when the effect B is greater than A).

The investigator is only interested in whether the new treatment is to be preferred

The example given earlier can be expanded if we assume the investigator is using an established drug, with effect B, in the treatment of hypertension and he is familiar with the effects of the drug and experienced in its use. However, he is willing to change to a new drug with effect A if A - B > 5 mmHg. If A - B is not greater than 5 mmHg the investigator wishes to continue with the established drug. The trial result must decide between two strategies: change to the new drug, and do not change. The investigator must be satisfied with this decision as the only outcome of the trial. The advantage will be a substantial reduction in the number of patients required for the trial. If the established treatment is better than the new treatment this fact may not be apparent as the numbers in the trial will not be sufficient to differentiate between the results A = B and B > A. This is a pragmatic decision making trial.

The investigator is certain that the control treatment cannot be better than the new treatment

In the above example, if the investigator decides not to change treatment, he should not be interested in whether the new treatment is the same as the control treatment or worse. However, when the investigator is interested both in whether A is greater than B and in whether A is similar to B, he must also consider the possibility that the effect B is greater than A. If it is not possible for B to be greater than A, then this outcome can be ignored with a reduction in the numbers required for the trial. In biological research, however, it is usually possible for A to be greater than B or B to be greater than A. An investigator using the new blood-pressure-lowering drug would certainly wish to report the fact that the new drug was worse than the old. He should not calculate the numbers required for the trial without taking account of this possibility.

The investigator must decide whether he wishes to make a yes/no decision or to quantify the differences between the treatment effects, A and B. If A is the treatment under trial and B the control treatment, the trial organiser must decide whether it is possible that B > A. If not a one sided test of significance may be employed (see below).

THE STATISTICAL SIGNIFICANCE OF THE DIFFERENCE TO BE DETECTED

The probability that an observed difference is due to the vagaries of chance is measured by a significance test to give a probability or P value. If the test gives a result less than five percent or one percent, then this is the probability of observing a difference as large or larger than the one observed, given that *in truth* no difference exists. In 100 trials this would occur 5 times with P = 0.05 and once with P = 0.01. The five percent or one percent levels are arbitrary cut-off levels at which we decide that the results are unlikely to have happened by chance and the result is statistically significant. These levels must be chosen during the design of the trial (so that the numbers required for the trial can be calculated) and constitute the first type of error, type I or α error. When the null hypothesis is true (that is, the treatment effects A and B are the same), the type I error is the probability of rejecting the null hypothesis when it is true. When calculating the numbers required for a trial, we assume that the desired effect is obtained and A - B = δ at a level of significance P where δ and P are defined in the objective.

Figure 6-1 shows the frequency of all possible results when the null hypothesis is true and A = B. The height of the graph represents the probability of a result, A -B = 0 being the most likely. The area under the graph is set equal to 100 percent and the right-hand and left-hand shaded areas represent the probability that a result will be obtained equal to or greater than both A - B = δ and B - A = δ respectively. The numbers for the trial are calculated so that, if P must equal five percent and B cannot be greater than A, the right=hand shaded area should represent five percent of the total area. This would happen with a one-tailed test of significance. However, in biological experimentation, B can usually be greater than A and to achieve a two-tailed probability of five percent, each shaded area must equal two and one-half percent as in this figure.

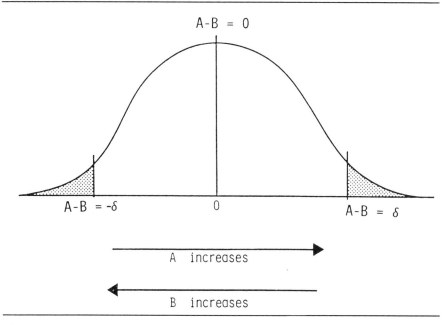

Figure 6-1. The graph represents the expected frequency of results when the effect of one treatment (A) equals the effect of a second treatment (B); that is, the null hypothesis is true. The shaded areas each equal two and one-half percent of the total area under the curve and a result greater than A - B = δ or less than A - B = -δ will differ from the expected zero result at the five percent level of significance.

THE POWER OF THE TRIAL

A negative result for a trial may be difficult to interpret. We must ask whether the number of subjects in the trial is sufficient to demonstrate an important difference between treatments should that difference truly be present. In many instances the numbers of subjects in a trial are so few that a negative conclusion is almost inevitable. Let us assume that 20 patients with coronary disease are treated for three months with an active antiplatelet drug and another 20 with placebo for the same period. It is unlikely that we will observe a coronary event in either group and if we conclude that treatment is not required for

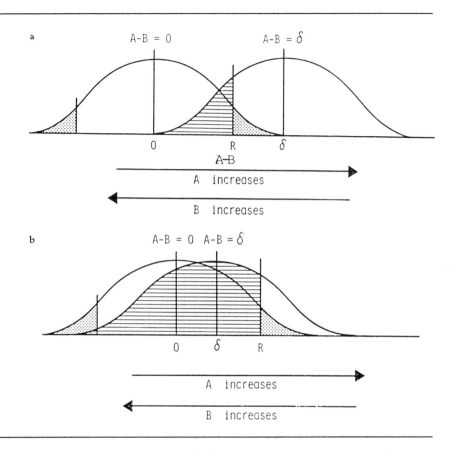

Figure 6-2. Two frequency distributions are represented when A - B is truly equal to zero and when A - B actually equals some difference to be detected, δ. The situation is represented in which there are sufficient numbers of subjects in the trial to ensure that a result R (just compatible with the null hypothesis) has a type II error of 10 percent. Figure b represents the situation in which there are so few patients in the trial that a result R (just compatible with the null hypothesis) has a type II error of 80 percent. ▦ type I error. ▤ type II error.

the prevention of ischaemic heart disease it would be obvious that our conclusion is lacking in power. Goodman and Berlin have defined power as *"the probability that, given a specified true difference between two groups, the quantitative results of a study will be deemed statistically significant"* [119]

The error in concluding that a given difference is not present when in reality it is, is known as the type II or ß error. The power of trial is the probability of avoiding ß error or

1 - ß. The numbers for the trial must be calculated so that the probability of a given result appearing compatible with the null hypothesis is small when a true difference of the specified size exists. Figure 6.2 gives an example of both adequate power (figure a, power 90%) and inadequate power (figure b, power 20%).

What power do we require in a trial?

Table 6-1 gives the results of antihypertensive treatment in four early placebo controlled trials. The first four rows give the effect of treatment on stroke incidence. In three trials treatment tended to reduce stroke incidence in male patients. There was no effect of treatment in women. One trial [121] did not observe a stroke event in either the actively treated or the control group; this trial was excluded from this section.

The final six rows of the table consider the effects of treatment on the incidence of myocardial infarction. There was not a significant reduction in myocardial infarction in any single trial, but in males the reductions were 100 percent, 10 percent, 100 percent and 100 percent respectively. The power of the trial, assuming that the actually observed difference was significant at the five percent level is reported together with numbers in the trial that would be needed to prove with 90 percent power that the observed difference was significant at this five percent level. Also given are the actual numbers enroled in the trials. The following points may be made:

1. In order to detect a large effect of treatment on the incidence of myocardial infarction, the numbers entered into the trial were only 45 percent, 0.4 percent, eight percent, and nine percent of the numbers required.

2. The power of these three trials was 42 percent, 8 percent, 17 percent, and 17 percent respectively.

3. The power of the trials that almost proved a reduction in stroke incidence was 75 percent and 55 percent respectively.

Table 6-1.   The results of four trials of active antihypertensive treatment versus placebos.

	Annual rate/1000		Difference (95% CI)	Significance of difference $(P_1 - P_2)$	Power of trial given $(P_1 - P_2)$ significant at 5% level (%)	Number in each group required to reach 5% significance, 90% power	Actual number in each group
	Control group (P_1)	Treatment group (P_2)					
STROKE							
Veterans Administration [64]	33	8	−25(−70,+20)	P = 0.30	23	360	72
Veterans Administration [52]	31	10	−21(−49,+8)	P = 0.14	75	340	190
Hamilton [120]							
Males	83	0	−83(−250,+84)	P = 0.34	55	35	11
Females	39	38	−1 (−120,+120)	P = 0.98	3	56,000	20
MYOCARDIAL INFARCTION							
Veterans Administration [64]	33	0	−33(−74,+9)	P = 0.10	42	160	72
Veterans Administration [52]	20	18	−2 (−29,+25)	P = 0.90	4	43,000	190
Hamilton [120]							
Males	21	0	−21(−140,+98)	P = 0.40	17	140	11
Females	26	13	−13(−97,+72)	P = 0.60	9	570	20
Cooperative randomised controlled trial [121]							
Males	21	0	−21(−77,+35)	P = 0.40	17	280	25
Females	16	13	−3 (−60,+54)	P = 0.90	3	15,000	33

Note: The effect of treatment on stroke and myocardial infarction rates is given together with the power of the trials, the number of patients who were in the trials, and the number who would be required to demonstrate the differences at the 5% level of significance and 90% power. CI = confidence interval (approx).

These calculations help us to get a feel for the power term that we might include in the calculation of the numbers required for a trial. Unlike the level of significance (type I error),there is less of a time-honoured tradition stating what is acceptable for type II or ß error. The current convention is 80% or preferably 90% [119] but a low-power term of 50 percent has been employed [122] as has a high level of 95 percent [123]. In a trial designed to make the decision between two treatments and not to determine whether the treatment effects are similar, a negative result will not be reported and therefore the power of the trial need not be considered as the trial result will only be "use one treatment" or "use the other treatment".

Confidence limits rather than power should be reported for negative results

Goodman and Berlin have stated *"A confidence interval can be thought of as the set of true but unknown differences that are statistically compatible with the observed difference"* and *"A 95% confidence interval is the result of a procedure that should include the true value 95% if the time"* not *"that any one confidence interval has a 95% chance of including the true value"* [119]. The table also gives these result for a probability of 95%.

Confidence intervals should be employed as they assess both clinical importance and statistical significance [124]. When a trial is reported it is not necessary to calculate power but confidence intervals must be reported. They are more easily understood than power terms.

Confidence limits provide a clear demonstration of the power of trials

Rose insisted that confidence limits should be reported for the results of any negative trial [125] and had in mind the use of 95 percent confidence limits, corresponding to the usual five percent test of significance. Baber and Lewis took up his point and, calculating less stringent 90 percent confidence limits, they showed that in 18 trials of the use of beta-blockers following myocardial infarction, the confidence limits encompassed a 50 percent reduction in mortality in 14 and an increased mortality (of any degree) in 16 [126]. 95% confidence limits are now generally applied.

VARIABILITY OF THE RESULTS

The number of subjects required for a trial will also depend on the variability of the end points being measured. Quantitative or qualitative measurements may have to be considered and their variability determined.

Quantitative data

When the trial is designed to detect changes in quantitative information, such as weight, blood sugar, or blood pressure, the variance of the measurement must be calculated. In a between-patient trial (two separate groups), the variance of the observation itself has usually been determined in preliminary studies and is calculated from the following formula:

$$\text{Variance} = \frac{\Sigma (x - \bar{x})^2}{n_a - 1} \quad \dots\dots\dots\dots\dots\dots\dots (6.1)$$

Where x is one of a number (n_a) of available and normally distributed observations (one per subject), $\bar{x}$ is the mean observation and $\Sigma(x - \bar{x})^2$ denotes the sum of the squared differences between individual observations and the mean. The end point of a trial may be an average of more than one measurement on an individual. For example, if three measurements of weight are averaged as an endpoint for a trial, the averages will be entered as the x's and the number, n_a, will equal the number of subjects.

For the within-patient cross-over trial or the change from baseline in a between-subject study, the variance of the change in the measurement must be calculated:

$$\text{Variance of change} = \frac{\Sigma \ (d - \bar{d})^2}{n_a - 1} \quad \dots\dots\dots\dots(6.2)$$

Where the notation is as for equation (6.1) except that d and $\bar{d}$ refer to the difference between two observations and not the observations themselves. The variance of a change may be considerably less than the variance of the original observations, thus reducing the numbers required for a cross-over trial (see below).

Qualitative or discrete data

Qualitative data include measurements such as the proportion of patients who die, improve, or relapse. The variance of such data depends on the proportion expected in the control and treatment groups. With quantitative data the variance of data in the control and treated groups is expected to be the same, but with qualitative data the variance will depend on the expected values.

$$\text{Variance in control group} = \frac{P_1 \ (1 - P_1)}{n} \quad \dots\dots(6.3)$$

where P_1 is the proportion with the end point in the control group and n the number of subjects that will be required in the control group.
  Similarly:

$$\text{Variance in the intervention group} = \frac{P_2 \ (1 - P_2)}{n}$$

where P_2 is the proportion with the end point in the intervention group and the same number of subjects (n) is assumed. To calculate the numbers for a trial with a qualitative end point the expected proportion with the end point in the control group (P_1) must be known from existing data; P_2 can then be calculated from the trial objective. For example, if the major objective is to reduce mortality by 50 percent, $P_2 = 0.5P_1$.

    With qualitative data, P_1 and P_2 are required
to calculate the number required for the trial
but the number of available subjects (n_a)
required for the original observation of P_1 is
not required.

CALCULATING THE NUMBERS

Some useful formulae will be presented but their
derivation is given in the statistical texts that
are referenced. We must first consider whether
the investigator wishes to estimate the effect of
the new treatment or to perform a less demanding
decision making trial.

The usual or explanatory trial

Schwartz and his colleagues considered the usual
trial to be an attempt to examine the magnitude
of treatment effects and to explain the
observations [118,127]. In figure 6-2 the
situations are illustrated when the control and
intervention group results are truly identical,
giving a difference in results of zero, and when
they are truly different by a difference δ. The
results of the trial will be distributed around
mean zero in the first instance and around mean
δ in the second instance giving the two frequency
distributions. The number of patients required
for the trial must be arranged so that the
difference designated as the objective of the
trial (R) is associated with given type I and
type II errors.

Type I error, α (two-tailed test)

In figure 6-2a, R is arranged to define two-and-
a-half percent of the area under both the right
and left sides of the A - B = 0 distribution and
the distance 0 to R is said to be 1.96
standardised normal deviates (1.96 SND).
    To calculate the numbers for a trial the
number of standardised normal deviates has to be
provided for a given α. Statistical textbooks
provide tables of these standardised normal
deviates, but when $\alpha=0.05$ (*two tailed*), SND=1.96.

Type II error, ß (power = 1 - ß)

In figure 6-2a if a just-significant result R or less is arranged to occupy ten percent of the area under the left side of the A - B = δ distribution curve (power 90 percent), then the corresponding standardised normal deviate (distance δR) will be 1.28.

Type III error, δ

A type III error occurs when we conclude that the truly better treatment is actually the worse treatment. In the explanatory trial type III error is vanishingly small.

Calculating the numbers for the classical explanatory trial

It will be assumed that either treatment may be superior giving a two-tailed test for α.

Quantitative data

Let n be the minimum number required in *each* of two groups.
Let d be the difference to be detected.
Let K be a constant equal to the square of the sum of the standardised normal deviate for α and ß $(SND_\alpha + SND_\beta)^2$.
Let s be the standard deviation of the measurement, then

$$n = \frac{2Ks^2}{d^2} \dots\dots\dots\dots\dots\dots\dots\dots\dots\dots\dots\dots\dots(6.4)$$

when α = 0.05 and ß = 0.10, $K = (1.96 + 1.28)^2$ = 10.5 as α is two-tailed and ß one-tailed.

Qualitative data

Let P_1 be the proportion having the event in the control group.
Let P_2 be the proportion in the intervention group.
Let n and K be as above.

$$n = \frac{K[P_1(1 - P_1) + P_2(1 - P_2)]}{(P_1 - P_2)^2} \quad \dots\dots\dots\dots(6.5)$$

This is an appropriate formula and a more exact formula is [128]:

$$n = \frac{(SND_\alpha \sqrt{[2P(1-P)]} + SND_\beta \sqrt{[P_1(1-P_1)+P_2(1-P_2)]})^2}{(P_1 - P_2)^2} \quad \dots(6.6)$$

Where $P = \frac{1}{2}(P_1 + P_2)$

The proportion of events in the experimental group can be adjusted for the number of dropouts to be expected from default or from withdrawal due to death from causes other than trial end points. Also, the numbers can be adjusted if the treatment takes a time to achieve full benefit [123]. Similarly the formulae may be adapted for more than two drug groups and for unequal allocation of patients between the groups (see below).

George and Desu [129] have also discussed the situation where survival times rather than events are to be compared.

The pragmatic decision-making trial

Schwartz and Léllouch named a decision-making trial a pragmatic trial [127] that is performed when we wish to make the decision between treatments A and B but do not care if they are truly similar. We must note the following:

1. The type I error (α), or the likelihood of saying one treatment is to be preferred when the treatments are equal is 100 percent.
2. The type II error (β), or the likelihood of saying that the treatments are equal when they are not, is zero.
3. The type III error (γ), the probability of preferring the worst treatment can be large and the trial numbers must be arranged to limit this error.

Table 6-2.   The three kinds of errors. Treatment effects are A and B. In reality the treatment effects may be equal (A - B = 0) or unequal B > A or A > B.

| | | Trial conclusion | | |
		B > A	A - B = 0	A > B
T	B > A	-	β	γ
R				
U	A - B = 0	$\frac{1}{2}\alpha$	-	$\frac{1}{2}\alpha$
T				
H	A > B	γ	β	-

α is the type I error; β the type II; and γ the type III error [118]. - = no error.

Table 6-2 illustrates the three kinds of error according to the observed results and the true results.

The calculation of the numbers required for a pragmatic trial is performed using formulae (6.4) and (6.5) above but K now equals the square of the standardised normal deviate for γ or type III error (one-tailed test) [118]. If γ is set at five percent, $K = 1.64^2 = 2.7$.

We can also decide on the use of a new drug when a new drug A is better than an established drug B by a certain numbers of units (D). The decision to use A could be made for results $> D$ and a decision in favour of B made when the result is less than D. The employment of a difference, D, in making a decision does not affect the numbers required for the trial [118]. The decision making trial would appear very suitable for the pharmaceutical industry when deciding whether or not to develop one drug in preference to another, or for health care managers when deciding whether to allow the costs of new treatments.

However, this type of trial design has not been adopted to any extent and trials described as 'pragmatic' are usually only testing overall treatment interventions without attention to which individual aspect of treatment actually produces benefit. An example is given by trials of health checks, the effect of the total package is assessed and the investigator is not certain which aspect of the package (the 'black box' effect) actually produces the benefit. In their design these trials do not employ the decision-making theory discussed above.

CALCULATION OF THE POWER OF A REPORTED TRIAL

The concept of the power of a significance test was discussed above. The power of a reported trial when the observed difference is assumed significant at the five percent level is given by SND_β when calculated by substitution of n and SND_α ($= 1.96$) in K and equations (6.4) and (6.5). This calculation of power led to the results reported in table 6-1. The power of a completed trial is not of interest when the trial has produced a positive results and Goodman and Berlin have argued that the calculation of power after a study is over is always inappropriate [119].

HOW THE SPECIFIED LEVELS OF ERROR INFLUENCE THE NUMBER OF PATIENTS REQUIRED FOR A TRIAL

The number of subjects required will be influenced by whether the trial is required for estimation of effect (explanatory trial) or as a mechanism for decision-making; by whether the α error is one-tailed, two-tailed or asymmetrically two-tailed; by the effect on α error of taking repeated looks at the data, and by the β error allowed. In addition we must examine the effect of factors influencing the level of variance of the data; the treatment effect that is to be detected; the effect of dropout and withdrawal; and whether or not we require equal numbers in the control and treatment groups.

Table 6-3.   Increase in numbers required for a trial with either α = 1% or 5% (α, two-tailed) and for increasing levels of power (1 - β).

	Increase in Numbers Required for a Trial				
1 - β	50%	75%	90%	95%	99%
α = 5%	1	1.8	2.8	3.4	4.8
α = 1%	1.7	2.8	3.9	4.6	6.3

The level of type I (α) and type II (β) error

Table 6-3 gives the increase in trial numbers that would be expected if α equals one percent and not five percent, and if 1 - β (power) equals 50 percent, 75 percent, 90 percent, 95 percent, and 99 percent. The baseline (number = 1) is assumed to be α = 5%, 1 - β = 50%. When α equals one percent and power 90 percent the trial numbers are increased fourfold and with a power of 99 percent the numbers are increased over sixfold.

One-tailed or asymmetric test instead of a two-tailed test for α

Schwartz and his colleagues have argued the case for the one-tailed test of significance [118] where the possibility that the intervention treatment is worse than the control treatment is ignored. However, we are usually equally interested in whether an intervention treatment is better or worse than a control treatment and a one-tailed test is not applicable. Schwartz and his co-authors pursued the argument and suggested that we may be interested in an unexpected finding but only if it is highly significant. They provided an example where the research worker was interested in a right-sided 2½ percent probability as in figure 6-1 but only in a 0.1 percent probability in the other direction. In their example the power of detecting a significant (P < 0.025) difference in one direction was 95 percent whereas the power of detecting a difference in the other direction was

70 percent. They stated, *"Thus we accept a rather large probability of failing to detect a difference on the left. This is quite reasonable; the test is primarily intended to detect a difference on the right ... A significant result on the left has to be considered as a byproduct."*

A by-product it may be, but although a trial of a treatment is usually mounted to detect a benefit, and an adverse effect of treatment is not anticipated (it would be unethical to perform the trial if it were), we must agree that the failure to detect an adverse effect may have the most serious consequences. If a trial lacks power in detecting a benefit, then patients who could be helped may not receive the treatment and this will be to their detriment. However, most trials are performed when a clinical impression suggests that the treatment is beneficial and one trial result showing no difference from, say placebo, will often result in the continued use of the drug. But a single trial showing a significant adverse effect can result in the treatment being withdrawn from use, used less frequently, or limited to certain patients. It can be argued that the detection of an adverse effect has more influence on patient care and is more important than the demonstration of a beneficial effect. It is only rarely that the investigator can justify an asymmetric test for type I error.

The effect of repeated looks at the data on the type I error

A statistical test, significant at the five percent level, indicates that the observed result has less than a five in 100 likelihood of having occurred by chance. However, the assumption is made that the statistical test is only performed once during the course of the trial. If the investigator makes the test after 20 patients have entered the trial, then after 40, 60, 80, etcetera, he will greatly increase the odds on reaching a five in 100 chance. Moreover he will presumably stop recruitment and end the trial when a significant result is observed. McPherson has pointed out that ten repeated tests on accumulating data at the one percent level of significance during a trial will be the same as an overall test for the trial at the five percent level of significance [130]. Similarly ten tests at the five percent level of significance will lead to an overall significance level test of 19 percent, almost a one in five probability of the finding being due to chance.

In a long-term trial, repeated testing may be necessary for ethical reasons. Table 6-4 gives the significance level for individual repeated tests (nominal levels of significance), one of which would have to be exceeded in order to achieve an overall level of significance of one percent or five percent. For example, if ten tests are to be performed and one gives a nominal level of significance of < 0.19%, then the true level of significance (P) is less than 1% (P<0.01). The maximum number of repeated assessments has to be decided at the planning stage of the trial and the nominal level of significance may be selected at that stage to give an overall level of significance of one percent or five percent. In chapter 12 we shall consider when a smaller level of nominal significance may be selected at the beginning of a trial, and larger levels later on, in order to prevent premature stopping of the trial. Nevertheless all the levels of nominal significance must accumulate to provide an overall α of 1% or 5%.

The nominal level of significance must be strictly adhered to. For example, with a maximum of ten tests, a nominal level of significance of one percent, and an overall level of five percent, the first test may be significant at the three percent level, but the trial cannot be stopped at this stage. The first statistical test

Table 6-4.   Nominal levels of significance (%) that must be exceeded if an overall level of significance is to be achieved of 1% or 5% when the test is repeated 5, 10, 15 , or 20 times [130].

		Number of Repeated Tests			
		5	10	15	20
Overall level of significance(%)	1	0.28	0.19	0.15	0.13
	5	1.59	1.07	0.86	0.75

is one of a series of ten and is as likely to be falsely significant as the last. Even though only one test has been performed and that was significant at the three percent level an overall level of significance of five percent cannot be claimed. The decision rule was to stop the trial if the one percent level is achieved in any one of ten tests and only if this level is reached in the first test may the trial be stopped.

If the investigator wishes to review the data constantly, perhaps with a view to an early completion of the trial, he may adopt a sequential trial design. This design is discussed in chapter 5 and takes into account the use of repeated significance tests.

HOW THE VARIANCE OF THE DATA INFLUENCES THE NUMBERS REQUIRED

Quantitative data

With quantitative data the variance of a result may be reduced by replication of the measurement. For example, it is reasonable to assume that the variance of fasting blood glucose will be reduced if it is measured every day for three days and the average taken as the end point of the trial, rather than if a single result is employed. The variance of an average result for each patient will be lower than that of a single measurement.

Qualitative data

With proportional data the variance is maximum at 0.50 and minimum at very high and very low values. More importantly, however, for a given proportional reduction, the higher the rate in the control group the greater is the effect of treatment in absolute terms. For example, a 50 percent reduction in a 90 per thousand event rate gives a fall in events to 45 per thousand (a difference of 45 per thousand) whereas a 50 percent reduction in a 20 per thousand event rate gives a fall to 10 per thousand and a difference

Table 6-5.   Numbers required for a trial to show that one antihypertensive drug reduces diastolic blood pressure by 10 mmHg more than a placebo.

End point	Between-Patient Study		Within-Patient Study
	Absolute values	Change in pressure	Change in pressure
Difference to be detected (mmHg)	10	10	10
Standard deviation (s and s_w)	13	8	8
Significance level	5%	5%	5%
Power	90%	90%	90%
K	10.5	10.5	10.5
Number of patients to be recruited to the trial	70	27	7
Number of observations required	70	54	14

Note: A between-patient design with and without baseline measurements is compared with a within-patient (cross-over) trial.

of only 10 per thousand. The higher the proportion in the control group getting an event, the lower will be the numbers required for the trial. This is one of the reasons for doing trials in selected high risk patients. However, Sondik and colleagues [131] considered a trial in which subjects with a high serum cholesterol were entered. The higher the serum cholesterol, the higher the risk. But if the subjects are to be detected by screening, the greater the serum cholesterol required to enter the trial the larger the number of subjects that have to be screened. In such circumstances it may be less expensive overall to enter medium-risk patients to a trial.

Reduction in variance by measuring a change from baseline

In certain circumstances, measuring change from baseline can drastically reduce the numbers required compared with a trial without baseline measurements. Table 6-5 gives the numbers required for a trial of a new antihypertensive drug. The standard deviation of between-subject blood pressure measurements (13 mmHg) was derived from a trial of such a drug and is given in column one. Similarly the standard deviation of within-patient changes in blood pressure is given in columns two and three (8 mmHg). The difference to be detected is 10 mmHg, α = five percent (two-tailed), β = ten percent. In a between-patient trial comparing two treatments without a baseline measurement, equation (6.4) estimates n = 35 so that 70 patients with 70 observations are required for the trial. With the same criteria but estimating change from baseline only 27 patients and 54 observations are required.

Reduction in numbers using the patient as his own control (a cross-over trial)

With the same parameters but a cross-over (within-patient) trial the formula is:

$$n_w = \frac{KS_w^2}{d^2} \dots\dots\dots\dots\dots\dots\dots\dots\dots\dots\dots\dots\dots (6.7)$$

where n_w is the number of subjects, $2n_w$ the number of observations, and S_w the standard deviation calculated from within-patient changes in blood pressure. In this example only seven patients giving 14 observations would be required for the trial but we have assumed that the difference between the treatments is the same in the first as the second treatment period. Cross-over (within-patient) trials were discussed further in chapter 5.

THE EFFECT OF ALTERING THE DIFFERENCES TO BE DETECTED BETWEEN CONTROL AND TREATED GROUPS

The smaller the difference to be detected the larger the numbers required for the trial. Table 6-6 gives the number of patients who might be required in a trial of a treatment to reduce the frequency of reinfarction in patients who have already suffered one myocardial infarction. It is assumed that the type I error is five percent and

Table 6-6.   Number of patients required for a trial of secondary prevention in myocardial infarction according to the event rate in the placebo group over the duration of the trial and the % reduction to be determined in the intervention group.

Event rate in placebo group/100	% reduction in events in the intervention group	Total number required for the trial
10	10%	29,460
10	20%	7,010
10	30%	2,960
10	40%	1,570
10	50%	950
20	10%	13,180
20	20%	3,160
20	30%	1,340
20	40%	720
20	40%	430

Type I error = 5%; power = 90%.

the power term 90 percent. The numbers are provided according to the rate in the placebo group over the duration of the trial and the expected reduction in events in the treated group. In this example 30 times as many patients are required to detect a ten percent reduction compared with a 50 percent reduction [123].

THE EFFECTS OF DROPOUT ON THE NUMBERS REQUIRED

Dropout is discussed in chapter 11 and is important when withdrawn subjects are excluded from the analysis. When they are retained in the analysis they may dilute any effect of treatment. Dropouts consist of those who default or otherwise do not follow the trial protocol and also consist of those who are withdrawn for criteria unrelated to the end points and those who are withdrawn from the trial for criteria possibly related to trial end points.

The Biometrics Research Branch, National Heart Institute [123] have published tables giving the numbers required for trial according to the expected dropout rate; these tables are appropriate when such subjects are to be excluded from the analyses.

DO WE NEED EQUAL CONTROL AND TREATMENT GROUPS?

Traditionally, treatment allocation has been arranged so that each group contains an equal number of subjects. However, two alternatives have been suggested, the use of 2:1 allocation of new: old treatments in trials comparing two treatments [132] and a relative increase in the number in the control group when several treatments have to be compared with this group.

A 2:1 allocation ratio with two treatment groups

A 2:1 allocation has been suggested for a comparison of a new treatment with an old or placebo treatment [132]. Less will be known about the effects of the new treatment and this is one reason for increasing the numbers receiving that treatment. A new treatment may also have to be compared in two different modes of administration or dose schedules, and one dose may be given to one-third of subjects, the second dose to another third and the old control treatment to the remainder, thus resulting in a 2:1 allocation. Peto and his colleagues [132] suggest that this strategy may allow different groups to participate in a trial of a new treatment even when they have divergent views on minor variants of treatment. Unequal allocation gives some loss of efficiency compared to a 1:1 allocation, but a 2:1 allocation is equivalent to performing a 1:1 allocation and eliminating about 10 percent of the patients from the trial. However, more unequal comparisons cannot be supported and 3:1 randomisation is equivalent to eliminating a quarter of the patients from the trial.

Unequal randomisation may also be employed when the costs of treatment vary. Cochran [133] and Nam [134] have discussed the square-root rule which states, *"If it costs r times as much to study a subject on treatment A than B then one should allocation $\sqrt{r}$ times as many patients to B than A."* This procedure minimises the cost of a trial while preserving power. Gail and colleagues [135] considered a similar situation where one treatment was more hazardous than the other and they also developed a case-saving rule.

A 2:1 allocation has great appeal when the new treatment may reduce mortality by 50%. If this proves to be true there will be as many deaths on the new treatment as the old treatment. When such a benefit is expected the 2:1 allocation may prove very efficient.

Unequal randomisation when comparing more than two groups

When more than one treatment group is to be compared with a standard control treatment, it may be desirable to increase the relative number receiving the control treatment. For example, the Coronary Drug Project trial compared five treatment groups with a placebo group. As each of the five drug groups was to be compared with the same placebo group, it was necessary to determine the final mortality in the placebo group with greater precision than in the actively treated groups (it was expected that the five-year mortality in the placebo group would be 30 percent and that active treatment would reduce this rate by a quarter). The Coronary Drug Project allocated 2.5 times as many patients to the placebo group (2,793 patients) as to any individual actively treated group (1,117 patients in each group) [136]. The ratio 2.5 : 1 was calculated by minimising the variance of the difference between the results in the drug groups and in the placebo group.

AIDS TO CALCULATING THE NUMBERS REQUIRED FOR A TRIAL

Tables have been published of the numbers required for a trial given the dropout rate, the difference to be detected, the event rate in the

control group, and the duration of the trial
[124]. Graphical representations of the numbers
required in each group according to the
percentage of patients expected to respond to two
treatments, and different levels of α error and
power have been published [122,137].

FAILURE TO PREDICT THE VARIANCE OF MEASUREMENTS IN THE TRIAL OR THE FREQUENCY OF EVENTS IN THE CONTROL GROUP

Calculating the numbers required for a trial
requires an accurate estimate either of the
variance of measurements used as the trial end
point or of the frequency of events occurring as
an end point.

An inaccurate estimate of the variance of measurements

If the variance of measurements during the trial
is lower than estimated, then less patients will
be required than calculated; if the variance is
larger, more patients will be required. Either
situation may occur: the initial estimate of
variance may have been lower if determined under
perfect standardised conditions that are not
reproduced in the trial; and the variance may
have been higher if calculated from observations
in normal clinical practice. In the trial,
standardisation and close attention to detail may
reduce the variability of the results.

An inaccurate estimate of the number of events in the control group

In a clinical trial, careful attention to the
patients' welfare may markedly reduce the number
of events that were expected in the control
group. This effect will be increased if patients
are withdrawn from the trial prior to an event
owing to an observed deterioration in their
condition. In the Hypertension Detection and
Follow-up Program trial [138] the patients who
were closely followed for hypertension showed a
reduction in mortality from several conditions

unrelated to hypertension, possibly the effect of an early detection of other disease processes.

It is possible, in a long-term trial, to observe more events in the placebo groups than expected. This can result from a failure to allow for the ageing of the population, In a ten-year trial of patients aged, say, 60, the correct number of events should be based on the frequency of events at age 65. If the number of expected events is based on the number at age 60, more events may be observed during the trial than expected owing to this error. However, the usual experience is that the events during the trial are 2-3 times lower than expected.

CONCLUSIONS

In order to calculate the numbers required for a randomised controlled trial, the major objective must be defined exactly. The treatment effects to be detected have to be designated and the distinction made between a trial designed to arrive at such an estimate (an explanatory trial) and one intended only to reach a decision (pragmatic decision making trial). The investigators must define the level of type I and type II error they will allow and also calculate the variance they expect in the trial end point. The methods of calculating the numbers required for a trial were given in this chapter together with the effect on these numbers of changes in the objective, level of type I and type II error, and change in variance of the trial end point. More specifically, the lower numbers required for the within-patient cross-over trial were discussed and the advantages and disadvantages of using unequal allocation procedures were given. The problems of a negative result were considered in some detail and the power of a trial defined and calculated. The concept of confidence limits was introduced and the problem reviewed when repeated looks at the data increased the level of type I error during the trial. Lastly, it was admitted that the assumptions made in the calculations may prove to be in error. However, it is very important to attempt to calculate the numbers required for a trial. Otherwise an inappropriate design and protocol may be adopted, leading to a predictably inconclusive result.

7. HOW TO ENSURE THAT THE RESULTS ARE FREE OF BIAS

Biased results are distorted and prejudiced in favour of one treatment or another. When discussing the assessment of a patient's progress in a trial, Bradford Hill [9] wrote. *"the judgements must be made without any possibility of bias, without any overcompensation for a possible bias, and without any possibility of accusation of bias."*

The results of a trial may be biased if an adequate control group is not provided and therefore the results cannot be interpreted; if patient allocation favours one group rather than another; when noncompliance influences the results; when patient or investigator bias affects the assessments; and when analytical bias distorts the presentation of the data.

AN ADEQUATE CONTROL GROUP MUST BE PROVIDED

Controls are necessary in clinical trials as otherwise any improvement or deterioration associated with giving the experimental treatment cannot be confidently attributed to the treatment. Randomised controlled trials rarely show as much benefit from treatment as most investigators assume will be the case. There are several reasons why an experimental treatment appears so much more effective in observational studies than in a controlled trial and these include the placebo effect, regression to the mean, and time trends in the condition being studied.

The placebo effect

In chapter 2 two early examples of placebo responses were reported; Haygarth's use of dummy wooden appliances to investigate the use of

Perkin's metal tractors which were supposed to cure by electricity [15] and Sutton's report of the effect of mint water in rheumatic fever [16].

In the 1920s the Western Electric Company carried out some experiments in its Hawthorne plant in Chicago. Illumination was either increased, decreased, or held constant. The workers were interviewed and the increased attention paid to them led to a rise in production, independent of changes in lighting intensity [139]. There was therefore an unplanned effect in the control groups and this became known as the Hawthorne effect. Ederer [140] considered that the placebo effect may arise either from a change in the social situation or from suggestion. Table 7-1 gives measurements of placebo effects mostly reported in a review by Beecher [141]. Even severe postoperative pain and angina can be satisfactorily relieved by placebo

Table 7-1. Placebo response in painful conditions and cough (largely derived from a table by Beecher [141]).

Complaint	Placebo	Percentage patients responding to placebo	Reference
Postoperative pain	Intravenous Saline	21 26	143 144
	Subcutaneous Saline	31 39	145 146
	Lactose by mouth	33	147
Angina pain	Tablets by mouth	26 38 38	148 149 150
Pain in myocardial infarction	Intravenous saline	54 (10 min) 31 (10-30 min)	151
Headache	Lactose by mouth	52	152
Cough	Lactose by mouth Subcutaneous saline	40 37	153 154

in up to 40 percent of patients, headache
alleviated in 52 percent, and cough in 37-40
percent.

Shapiro [142] has concluded, *"we are led to the
conclusion that the history of medical treatment can be
characterized as the history of the placebo effect, since
almost all medications until relatively recently were
placebo."* Dollery [155] suggested that the placebo
effect can be both psychological and physical. He
considered that *"A friendly smiling face, comforting
personality and a caring environment make most people feel
better. The effect is reinforced if the caring environment
is combined with the powerful suggestion that a specific
treatment will be of benefit."* However, Dollery
proceeded to point out that physical changes can
result from suggestion. He commented that
hypnosis can lower airways resistance in some
patients with asthma and that a placebo may
stimulate the release of encephalins in the brain
and thereby have a true pain alleviating effect.
Encephalins act on opiate receptors in the brain
and when naloxone, a drug that blocks these
receptors, was compared with placebo to assess
the pain relieving effect of this drug, it was
found that placebo was more effective. As a
placebo can be so powerful, it is usually
necessary to employ a treatment, placebo or
otherwise, in the control group so that the
effect of the experimental medicine can be shown
to be greater than the control treatment. The
method of employing placebo treatment is
discussed below.

Regression to the mean

In a medical practice, patients will usually be
seen when they have developed symptoms. If the
patients are suffering from a self-limiting
illness, then they can be expected to improve and
any treatment prescribed may be given the credit
for this improvement. Similarly a patient may
have a chronic condition that varies in severity
so that he or she will see the doctor when the
disease is most troublesome and then be expected
to improve. Again, treatment may receive the
credit. The phenomenon of selecting a group of

patients when they have a high recording of some measurements (for example, number of symptoms or blood pressure) and then observing an overall improvement is one form of regression to the mean (see also chapters 11 and 14).

Blood pressure varies considerably and if a group is selected when the measurement is high, it can be expected that next time the average pressure will be lower. Similarly if a group is selected because the reading is low, the subsequent mean reading will be higher. Regression to the mean is the movement of a result from an outlying position nearer to a more moderate value as a consequence of an initial selection of the subjects.

Time trends

When historical controls are employed, a reduction in the severity of disease with time may bias the interpretation of results in a group being studied later. This problem has been reviewed by Peto and his colleagues [132] who concluded that the later group may also be affected by changing referral patterns, changes in ancillary treatment, and the selection of patients. They point out that ignoring a few old or untreatable patients in the new series may make a large difference. The content of the new group is obviously open to manipulation, however Peto and co-workers considered that historical data may be useful in assessing the results of randomised trials. For example, if randomised controls have fared worse or better than expected from previous studies it may cast doubt on the trial conclusions. The control group in the University Group Diabetes Program trial did better than expected but this criticism has not been very persuasive (chapter 19). On the other hand, the error from using a historical control as the only control has been conclusively demonstrated [108, 132, 156, 157].

BIAS DUE TO AN UNEVEN ALLOCATION TO THE TREATMENT GROUPS

It is essential that the treated and control patients are similar in order that any differences in outcome can be attributed to the treatment and not to other factors. Thus far everyone agrees, but how to obtain similar groups is open to some discussion. In this section we discuss the two classical methods: (1) using the patient as his own control (cross-over studies) and (2) random (chance) allocation of patients to distinct and concurrently treated groups. Randomisation is expected to result in the groups being similar. The futility of using historical controls is discussed above and this section deals mainly with the advantages, disadvantages, and methods of randomisation.

The patient as his own control

When a patient is given one treatment and then a second the patient acts as his own control. The trial is known as a cross-over trial and the design is discussed in chapter 5. The order in which the treatment is given may be important if the baseline is changing with time, and randomisation should be employed to ensure that the patients who receive a certain treatment first are similar to those who initially receive a different treatment. A cross-over design is only feasible for a chronic condition that reverts to its original state with the cessation of treatment (for example, high blood pressure or diabetes mellitus).

Randomisation

The technique of randomisation in clinical trials refers to the *"assignment of treatments to patients using a chance procedure"* [158]. Randomisation eliminates bias in patient allocation, renders the treatment groups equal in most important respects, ensures statistical tests are valid, and improves the acceptance of results. It also has some potential disadvantages.

Advantages of randomisation

i)   Treatment allocation is free of bias

If an investigator allocates patients to either
of two treatments and not at random, he may do so
on certain criteria. For example, he may allocate
on the basis of the severity of disease. In
chapter 2 we discussed the fact that James Lind
allocated patients with the most severe scurvy to
a particular treatment. The results in such a
severely affected group may be biased against
Lind's selected and possibly favourite treatment.
A more recent example was given by a trial of
anticoagulant therapy in myocardial infarction
[159] conducted by the Committee on
Anticoagulants of the American Heart Association.
The investigators considered that there was
sufficient doubt as to the efficacy of
anticoagulant therapy to warrant a clinical trial
of this therapy against the usual basic medical
care. Unfortunately, they did not randomise the
patients but allocated those admitted to hospital
on even days to the control group and those
admitted on odd days to the anticoagulant group.
The results were published and the patients given
anticoagulants fared much better than the
controls. However, it was noticed that many more
patients were allocated to anticoagulant therapy
than to control treatment (a ratio of 1.4:1,
table 7-2).
    It appeared that some referring doctors
favoured anticoagulants, and if a patient had a
myocardial infarction on an even day they
preferred to admit them to hospital on an odd
day. Seriously ill patients could not wait and
had to be admitted on an even day giving a
smaller control group with severe disease who did
badly.
    Problems were not limited to recruitment in
this trial and difficulties arose both in
performance and analysis. Patients in the control
group could be given anticoagulants at the
request of a private physician and 31 such
patients were transferred to the treated group.

Table 7-2. The number of episodes of myocardial infarction involved in the American Heart Association trial of anticoagulants [159].

	Excluded	Control	Treated
Admitted on even days (total 490)			
1. Not given anticoagulants		395	
2. Given a/c's for complications		35	
3. Given a/c's to prevent complications			31
4. Excluded	29		
Admitted on odd days (total 604)			
1. Given anticoagulants			546
2. Not given a/c's Miscellaneous reasons		12	
3. Not given a/c's Contraindications			12
4. Excluded	34		
Total	63	442	589

Note: Twelve patients had two admissions and one had three. The number of patients involved in the trial was therefore 1,080.

Moreover, 35 patients who developed thromboembolic disease had to receive anticoagulants but remained in the control group. If the protocol cannot be followed after entry to the trial and the patient is withdrawn or transferred to another treatment group, then the patients should remain in their original randomisation group in order to avoid bias (see the intention-to-treat principle, chapter 14). This rule was followed for 35 patients given anticoagulants in the control group and 12 not given anticoagulants owing to contraindications in the treated group. However, the 31 episodes of active treatment in the control group that were discussed earlier and the 12 episodes of control treatment in the treated group were not analysed on the intention-to-treat principle. Lastly, 12 patients had contraindications to anticoagulant therapy and should not have entered the trial. We should also refer to episodes rather than patients when describing the results of this trial as 13 patients were allowed to enter the trial more than once. This duplication should not be allowed (chapter 14).

The benefit from treatment reported from this trial was a 27 percent reduction of mortality in men and a 36 percent reduction in women. The greater benefit in women does not agree with other studies and the ratio of treated to controls entered in this trial was 1.5:1 for women and 1.3:1 for men suggesting that the failure to randomise may have distorted results more for women than for men.

ii)   The treatment groups are similar with respect to important confounding variables

Confounding variables are those factors other than treatment that are related to both treatment and outcome. They are otherwise known as interfering or nuisance variables and the effect of such variables is to confound (alter) the effect of treatment. If, for example, survival from cardiovascular disease is being examined according to whether or not the patients are given a lipid-lowering drug or placebo, it is important that the active and placebo groups are similar for the known risk factors for cardiovascular disease such as high blood pressure or cigarette smoking. Randomisation may be expected to lead to two groups similar with respect to these characteristics.

If the groups are similar, it does not imply that they are identical. However, it can be anticipated that the two groups, for a given confounding variable, will not differ at the 5 percent level of significance with respect to this item. When considering a large number of confounding variables it is unlikely that randomisation will ensure that the groups are similar in every respect; in fact, with 14 variables there is a greater than even chance of one item differing significantly at the 5 percent level. But most confounding variables will not differ with statistical significance between the two groups. On the other hand, nonsignificant differences may still be of biological importance. This is discussed below.

iii) Randomisation ensures that many statistical tests are valid

Many statistical tests assume that the populations to be compared are randomly drawn from a larger single population. This is manifestly true for randomised controlled clinical trials where the treatment groups are randomly drawn from the total group of patients.

iv) Randomisation improves the chance that the results will be accepted

Weinstein [160] has pointed out *"one randomised trial with 100 patients can dramatically change physician behaviour, whereas the experience of 100,000 patients might be neglected."* The clinician will naturally be influenced more by an objective unbiased study such as the randomised controlled trial. When randomisation has not been carried out but the data adjusted afterwards to correct for differences between the groups, Weinstein went on to say, *"Although matching or covariance analysis may satisfy the investigator himself that all important nuisance variables have been washed out of the analysis, there will always be someone somewhere who will complain that the analysis did not control for colour of eyes or sunspots ... that he believes ... to be confounding the results."* Randomisation usually provides groups similar with respect to the colour of their eyes – and sunspots!

Disadvantages of randomisation

i) Randomisation may lead to unequal numbers in each group

When large numbers of patients have to be randomised, the numbers in each group are likely to be very similar. With small numbers (as in a small trial or in one centre of a multicentre trial), the number receiving one treatment may differ considerably from those receiving a second treatment. Zelen [158] gives an example for 24 patients where random allocation led to nine on one treatment and 15 on another. This problem may be overcome by restricted randomisation which is discussed below.

ii) Randomisation may still provide groups that differ in some important respect

After randomisation, bad luck may lead to the groups being different for an important variable and such a difference may be both statistically and biologically important. With small numbers the differences may not be statistically significant, but they may still be of biological importance. Considering the previous example of 24 patients, it is possible that even if 12 patients received one and 12 the other drug, eight of one group but only four of the other group may be male. This difference would not reach the five percent level of statistical significance but the two groups could hardly be said to be similar with respect to the proportion of men in each group.

Important confounding variables can be allowed for prospectively by stratification (that is, randomisation within certain strata such as age groups) (see below). Retrospectively an imbalance can be allowed for in the analysis [156].

iii) Randomisation alone may be less efficient than matching

Weinstein [160] stated that *"If a confounding factor is known to be present, a smaller sample size would be sufficient to achieve comparable statistical information content by matching or adjustment procedures than by randomisation alone."* But attempts to get matched patient pairs may have two undesirable consequences. First, as the trial proceeds it may become apparent that another important confounding variable should be considered and the pairs will not be matched for this variable; and second, partners may not be found for certain patients and they may have to be omitted from the trial. Randomisation within a limited number of strata may provide similar groups when more than a few subjects are to be entered.

iv)   Is it desirable for some patients to be allocated to an inferior treatment by chance?

The result of the trial may show that one treatment is inferior and it has been suggested it is immoral to decide this outcome by chance. Randomisation can be adapted to limit the number of patients who receive an inferior treatment (chapter 5). But if two treatments are not equal some patients must suffer. A superficially attractive way around this problem is to ask the patients to select their treatment. This is usually not appropriate for drug treatment but has been suggested for operative treatment [160]. For example, one operation may be thought to have a high initial mortality but an increased likelihood of long-term survival and a second operation may have a lower initial mortality but worse long-term survival. A young patient with a family may well opt for the second operation. As in this example, it is very unlikely that the patients who opt for one operation will be similar to those who choose the other operation, leading to the problem that the effect of treatment will be confounded (confused) with the different characteristics of the patients. Randomisation remains the only safe method of selecting the two groups and when the investigator is in genuine doubt concerning the preferred treatment, only chance allocation can be ethical.

METHOD OF RANDOM ALLOCATION

We must discuss at what stage of the trial we should randomise, how to use random number tables for randomisation, whether to restrict randomisation to give an equal allocation of subjects between treatment groups, whether to choose an unequal allocation, and lastly whether or not to randomise within certain groupings or strata.

The stage of the trial at which a patient should be randomised

In clinical trials the investigator should first decide that a patient is eligible for the trial, then randomly allocate the patient to a treatment group. He must not be aware of the next treatment to be allocated as with this knowledge, bias in allocation may occur. For example, if the investigator is aware that the next patient to be entered will receive placebo treatment, he may be unwilling to enter a patient with severe disease. The placebo group will then contain an excess of less severely affected patients. The investigator must be certain that the patient is suitable for any of the trial treatments and then determine the result of randomisation.

Method of determining the result of randomisation

To ensure that the investigator is not aware of the next treatment to be allocated he may be required to contact a central coordinating office to determine the result of randomisation. Alternatively, if the result of randomisation is held in the investigator's office, the treatment to be allocated should be held in sealed numbered envelopes, opaque when held up to the light, to be opened sequentially as required, only after the patient's name and other details have been written on the appropriate envelope and the envelope must be signed in the presence of a witness who also signs.

Method of simple randomisation

When there are only two treatments, randomisation can be achieved by simply tossing a coin. However, it is usually administratively more simple, and less open to manipulation, to decide the randomisation in advance. This can be done conveniently using a table of random numbers as illustrated by table 7-3, the four-digit random numbers having been generated by a computer program. The table may be read as four-digit

Table 7-3. Four digit, computer-generated random numbers

0011	0858	4371	7901	2691
1893	2018	1425	6942	0755
6658	9411	5940	5568	7667
5147	6128	9689	6802	0238
7886	1512	3488	1699	7461
9675	5715	8699	2543	0769
7756	1808	5655	4214	6091
6931	9953	6577	3369	2761
7134	5116	3817	6314	3651
3921	4119	9007	5648	3664
4893	7074	0631	7365	2477
5848	8690	1833	7555	6690
1849	8756	6811	0913	8525
9462	9739	0522	9183	8797
9843	8360	7114	3640	3097
2322	4221	6663	9846	8555
1705	7965	7440	8084	6412
0987	4087	6593	4548	1165
7453	9080	1084	1354	1610

numbers or single figures or any number of digits either horizontally or vertically. Table 7-4 gives the results of using single numbers horizontally and allocating treatment A when the random number is even and treatment B when the number is odd. The starting number was selected at random in the first row, the sixth number from the left. After 20 patients the ratio of patients on A:B was 1:1.5 and after 100 patients, 1:1. With a small trial grossly unequal numbers of patients may be allocated to two treatments. This will not be a problem with large trials unless they are multicentre, where the small numbers allocated to individual centres may result in some centres having a markedly unequal distribution of treatments. The problem of unequal treatment groups is overcome by restricted randomisation [140], otherwise known as block randomisation [158].

Table 7-4. Allocation of two treatments, A and B, to 100 patients according to whether the random numbers in table 7-3 are even (A allocated) or odd (B allocated).

Patient	Random Number	Treatment A or B	Cumulative number on A	B
1	8	A	1	0
2	5	B	1	1
3	8	A	2	1
4	4	A	3	1
5	3	B	3	2
6	7	B	3	3
7	1	B	3	4
8	7	B	3	5
9	9	B	3	6
10	0	A	4	6
11	1	B	4	7
12	2	A	5	7
13	6	A	6	7
14	9	B	6	8
15	1	B	6	9
16	1	B	6	10
17	8	A	7	10
18	9	B	7	11
19	3	B	7	12
20	2	A	8	12
.	.	.	.	.
.	.	.	.	.
.	.	.	.	.
40	9	B	19	21
.	.	.	.	.
.	.	.	.	.
.	.	.	.	.
60	6	A	28	32
.	.	.	.	.
.	.	.	.	.
.	.	.	.	.
80	1	B	42	38
.	.	.	.	.
.	.	.	.	.
.	.	.	.	.
100	5	B	50	50

Note: Single numbers have been selected working horizontally and starting at the first line (see text).

Restricted or block randomisation

Restricted randomisation ensures that within a block of patients equal numbers are allocated to each treatment. For example, when it is decided to restrict the randomisation so that every group of four patients includes two on treatment A and two on B, then Zelen [158] suggested allocating a number to each of the six ways that the treatments can be allocated in groups of four. As before, the random number table is consulted, not to allocate a single patient to a single treatment, but to allocate a sequence of four patients to four treatments as in table 7-5.

Table 7-5. The use of a random number table to achieve restricted randomisation

Random number	Sequence
1	A B A B
2	A A B B
3	B B A A
4	B A A B
5	A B B A
6	B A B A

Note: Randomisation is restricted to give two patients on drug A and two on drug B for every four patients entered into the trial. The random number is selected from a table of random numbers and gives the treatment allocation for the next four patients to be entered into the trial.

The disadvantages of restricted randomisation is that the investigator may be able to predict the next treatment to be allocated. For example, an investigator who is aware that randomisation is restricted for blocks of four subjects would be able to predict the allocation for the fourth, eight, or twelfth patient, and so on. With a multicentre study, restricted randomisation at the coordinating centre may lead to balanced numbers in the trial as a whole but not for an individual participating centre and the local investigator would be unable to predict the treatment. However, it is desirable to have roughly equal numbers at each centre as patients may do particularly well, or badly, at one centre. If a certain centre has a predominance of one treatment the result may appear to be due to that particular treatment. It is therefore often necessary to randomise within each centre to ensure that a roughly balanced allocation occurs at the centres. Zelen has described a method, called balanced block randomisation [158], which gives a balanced allocation for each centre while making it difficult to predict the treatment sequence. The method requires the use of an auxiliary random number table whereas variable block allocation can achieve the same ends and be more easily understood. Variable block allocation is therefore described here.

Variable block randomisation

In variable block randomisation the investigator does not know the number of patients to be recruited before balance is achieved. Equal numbers may be reached after, say, four or six patients and this makes it difficult for the investigator to predict the next treatment to be allocated. Balance after four to six subjects can be achieved by extending table 7-5 so that random numbers seven to 26 cover the 20 different sequences for blocks of six (table 7-6). The random numbers in table 7-3 must be examined in pairs and four block or six block sequences allocated according to numbers 01, 02, ... 26. A further alternative to variable block randomisation is adaptive randomisation. This is a complex procedure whereby the probability of selecting a treatment can be reduced if an excess of that treatment has been allocated [158].

Randomisation within strata

In the same way as restricted randomisation prevents a disproportionate number of patients being allocated to one treatment, stratification ensures that one treatment group does not include an excess or deficit of subjects with a certain confounding characteristic (for example being male). Inequality between groups is usually only a problem when the numbers in a trial are small. However, most investigators would stratify: (1)

Table 7-6. The 20 different possible sequences for two treatments and blocks of six patients

Allocation number	Treatment sequence
07	A A A B B B
08	A A B A B B
09	A A B B A B
10	A A B B B A
11	A B A A B B
12	A B A B A B
13	A B A B B A
14	A B B A A B
15	A B B B A A
16	A B B A B A
17	B B B A A A
18	B B A B A A
19	B B A A B A
20	B B A A A B
21	B A B B A A
22	B A B A B A
23	B A B A A B
24	B A A B B A
25	B A A A B B
26	B A A B A B

Note: The random number can be searched for in random number tables to decide on the treatment allocation. Random numbers 1-6 can be used to allocate treatment to blocks of four subjects (table 7-5), providing variable block randomisation.

by centre (in a multicentre trial); and (2) by gender.

Randomisation is carried out within each centre and for both sexes separately. Weinstein [160] has argued that *"matching, blocking or adjusting may be far more efficient devices than purely randomising. Why let chance do what one can do for oneself?"* Matching can be considered to be the most extreme form of stratification with the subjects in matched pairs being randomly allocated, one to receive one treatment and one the other. Retrospective matching or adjusting the data is less desirable than starting with similar groups.

Peto and associates [132,161] considered stratified allocation to be an unnecessary complication for large trials. In such trials chance will usually ensure that the treatment groups are comparable and retrospective stratification can be used to compare the treatment effect in one stratum (for example, males) with the effect in a second stratum (in this example, females). Stratification or matching cannot ensure that the treatment groups are equal in all important respects. Often the variables that are known to be of importance are too numerous and stratification for these features would lead to several strata containing too few patients. Also, if previously unrecognised but important features are discovered during the course of the trial or during analysis, stratification or matching will not have coped with these. In conclusion, the number of strata should be kept at a minimum and restricted randomisation may be carried out within each strata if desired.

Randomisation schedules that do not allocate treatments to an equal number of patients

When comparing a new treatment with established therapy, Peto [161] has suggested that the numbers allocated to a new treatment may be increased by giving two patients the new treatment for every one patient given the control treatment. He states, *"there will be an unbiased randomised comparison which is very nearly as sensitive as an ordinary equal-groups randomised trial ..."* In long-term trials of survival, when the new treatment is expected to reduce mortality by 50 percent, then 2:1 randomisation may be expected to yield an equal number of deaths in two groups. Two-to-one randomisation can be achieved by using twice as many random numbers for allocation to the new treatment as are used to allocate to the control treatment. Unequal randomisation is very attractive when little is known about the new treatment whereas a good deal is known about the old or control treatment.

Randomisation of the group rather than the individual

In trials where the individuals are randomised to receive a certain treatment or advice and they are aware of the nature of this 'treatment', they may discuss their 'treatment' with others. If these second parties constitute the control group and thereafter change their behaviour or request the treatment, *contamination* is said to occur. The possibility of contamination was foreseen in the WHO factories study [162] where the effect of lifestyle changes was determined in the prevention of cardiovascular disease. If individuals in a factory were to be randomised to receive advice on stopping smoking, limiting saturated fat etc, it was expected that they would convey this good advice to their colleagues who constitute the control group. Thus the effects of the intervention may be diluted by contamination of the control group if that group were to adopt the intervention advice. Individual factory workers were therefore not randomised but all workers in a factory were randomised to receive or not receive the intervention. The factories were randomised, not the individuals. Similarly in the MRC study of Assessment and Management in the Elderly, currently underway in the United Kingdom, two different levels of screening activity are compared (minimal and intensive) and two methods of assessing problems detected by screening (by a team based in General Practice (primary care) or by a Geriatric Evaluation and Management team based in a hospital out-patient environment). There was concern that in a particular general practice, individuals randomised to the control groups may insist on the more intensive form of screening or assessment when they become aware of these facilities. It was therefore decided not to individually randomise the 50,000 subjects involved but the 108 general practices instead. Within a single general practice all subjects receive the same type of intervention, limiting contamination of the 'control' groups.

ALTERNATIVES TO RANDOMISATION

There is usually no good alternative to randomisation. The following section gives some alternatives that have been suggested and the problems that may be encountered with them.

i)   Allocation according to date of entry to the trial

This was used in the trial discussed above [159]. The allocation was known in advance, odd dates for one treatment, even dates for another, and the trial entry was manipulated to give unequal groups. Prior random allocation, using random number tables, is simple and not open to such manipulation.

ii)   Allocation according to the hospital number

The hospital number has been employed, even numbers being allocated to one treatment, odd numbers to the other. This can be open to manipulation as with the date of entry to the trial.

iii) Allocation according to the initial letter of the subject's name

If the initial of the subject's name is A to M, he or she can be allocated to one treatment, and if it is N to Z to the other. This is a very unsatisfactory method. Not only can allocation be manipulated but the treatment groups may be very different. For example, in the United Kingdom there are more patients with surnames starting A to M than N to Z. Moreover, A to M will include an excess of Scotsmen with names starting with Mc; N to Z will include a greater proportion of Irishmen with names starting O' and also Asians with surname Singh or Patel.

iv)   Allocation according to the wishes of the patient

This procedure has been proposed in the context of trials of surgical procedures [160]. Although it is desirable to conform with the patient's wishes, it is both possible and probable that patients choosing one form of treatment will differ from those choosing another. Obvious differences can be adjusted for retrospectively in the analysis, but differences may remain that are not recognised and can be confused with treatment effects. Random allocation is by far the safest procedure and cannot be recommended too strongly.

v) *Allocation according to the preceding results*

An example of such a strategy is given by the play-the-winner rule: if a treatment is followed by success the next patient also receives this treatment; if a treatment results in failure the next patient receives the alternative treatment. Play-the-winner rules are discussed in chapter 5. Not only must the results be known quickly but the allocation may be open to manipulation.

Conclusions on randomisation

It cannot be stressed too strongly that randomisation is necessary to ensure that the different treatment groups are similar in most respects. Occasionally bad luck will still lead to unequal groups and retrospective adjustment of the data becomes necessary. However, randomisation is the best safeguard that the groups are equal and that the investigator does not consciously or unconsciously manipulate entry to the trial, thereby producing groups of unequal numbers of differing confounding factors.

A subject must be eligible for the trial and *then* randomised to a treatment group. Randomisation may be restricted to give equal numbers in each treatment group and restricted randomisation may be employed within strata to ensure equality of confounding variables. Randomisation can be employed when there are more than two treatment groups and when more patients are to be allocated to one treatment group than another. In a within-patient cross-over trial, randomisation is also required to determine the order of treatment and to make sure that whether treatment A precedes treatment B, or vice versa, is a matter of chance and not open to manipulation by the investigator.

ALTERED COMPLIANCE MAY BIAS THE RESULTS

Compliance may be defined as adherence to therapeutic advice and it may radically alter the interpretation of a trial result. Figure 7-1 illustrates two possible assessments of the effect of stopping smoking on total mortality in men; the first arose from an observational study of the difference between subjects who stopped smoking and those who continued and the second assessment resulted from a randomised controlled trial. The observational study [163] reported an excess of mortality in those who continued to smoke compared with those who spontaneously stopped smoking (0.15 percent/year). The observational study suggested a benefit from stopping smoking whereas the randomised controlled trial revealed a slight and nonsignificant increase in total mortality in those who were given antismoking advice [164]. The result of the trial did not fulfil the hopes raised by the observational study. A longer period of observation may reveal different results, especially as the adverse effects of smoking, if reversible, will take time to be corrected.

A further reason for the poor result may have been noncompliance with the therapeutic advice. Not all the intervention group stopped smoking (only 36 percent had stopped three years after the advice was given). Also some of the control group did stop smoking (14 percent after three years). However, these facts may not explain all the differences between the results of the observational and intervention studies. In an intervention study subjects may be persuaded to stop smoking who would not otherwise do so. In the observational study, those who stopped smoking were presumably a health-conscious group who may also have been moderate in their diets and taken exercise. Not surprisingly these men had a low mortality and it must be noted that subjects who never smoked had the lowest mortality. The randomised controlled trial was the best test of the effect of intervention and tested the strategy of giving antismoking advice. The strategy had only a modest success, the

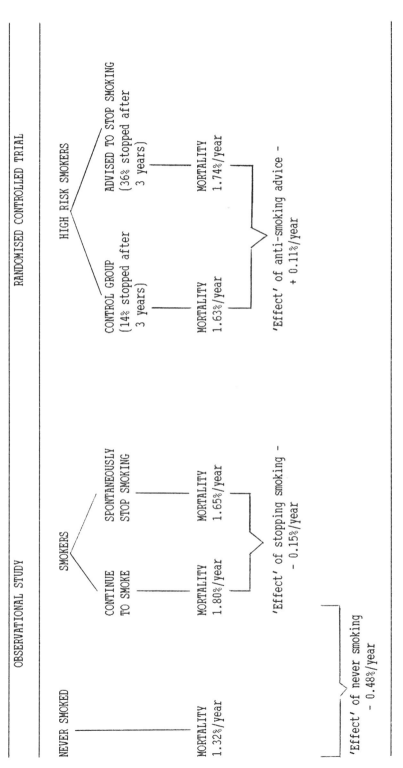

Figure 7-1. The effects of stopping smoking as determined from an observational study and a randomised controlled trial (see text).

majority of subjects being noncompliant with the therapeutic advice. In this example, noncompliance can be expected to result in a smaller effect of intervention than predicted from observing only subjects who stopped smoking spontaneously.

It would appear attractive to assess the effect of smoking in the trial, only in those who take the advice; this is equivalent to considering only those who are compliant with treatment. This analysis would be appropriate if the trial tests only for a relationship between stopping smoking and lower mortality. However, the experimental treatment in the trial was giving anti-smoking advice, expecting to demonstrate a causal association that could be reversed. If only those who stopped smoking were analysed, the results would apply to men who were relatively health-conscious: the same group that was assessed in the observational study. The effect of giving anti-smoking advice was determined without bias in this trial but the *interpretation* must include an appreciation of the degree of *compliance* in the intervention group and *contamination* in the control group.

The results of a randomised controlled trial may be worse than expected when noncompliance is a problem or better than expected when compliance is improved under trial conditions. The results of a trial must be analysed primarily on the intention-to-treat principle (chapter 14) in order to avoid bias from selecting a group with an increased compliance.

PATIENT BIAS MUST NOT INFLUENCE THE RESULTS

We have discussed the strength of the placebo response. During treatment the patients wish to feel better and will tend to report favourably, implying an improvement in their clinical condition and also pleasing the investigator to whom they may feel indebted. Ederer [140] quoted Francis Bacon: *"For what a man had rather were true he more readily believes."* Shapiro [142] stated, *"The desire for successful outcome is felt so strongly in both patient and investigator that objectivity cannot be guaranteed. Both have an emotional stake, overt or occult, in the result. Further, the giving of any treatment especially by needle injection, is a strong psychotherapeutic stimulus in itself."*

An adequate control group has to be employed to measure any placebo effect. However, the placebo effect will almost certainly not be observed if the patient knows that a particular treatment is a dummy. Therefore the patient must not know which treatment he is receiving and he is said to be blind with respect to the treatment. When the patient is not aware of the nature of his treatment but the investigator does know, a trial is said to be single-blind. When both the patient and the investigator are unaware as to the treatment being given, the trial is said to be double-blind. Investigators in ophthalmological research prefer the term *doubly masked* owing to the disagreeable association and possible confusion that may result from the use of the term *blind* [165]. The term *treble blindness* has also been advocated and is used when an analysis is performed with the report writer being unaware of the treatment allocated to the group in an otherwise double-blind trial. The strategy sometimes proves very useful when a committee is monitoring the progress of a trial and performing interim analyses (chapter 12), or when a group is writing the first draft of a trial report. If a monitoring committee favours one treatment, being uncertain of the nature of the treatment leading to the results may reduce a desire to stop the trial. Similarly if the writing group favours one treatment then a tendency to exaggerate the advantages of this particular treatment may be reduced when the 'best' treatment may prove to be the treatment which was originally less favoured.

INVESTIGATOR BIAS MUST NOT INFLUENCE THE RESULTS

Few would argue that the patient is an unbiased observer of his well-being, the placebo effect being too well recognised. However, we must discuss how difficult it may be for an investigator to make unbiased observations.

138

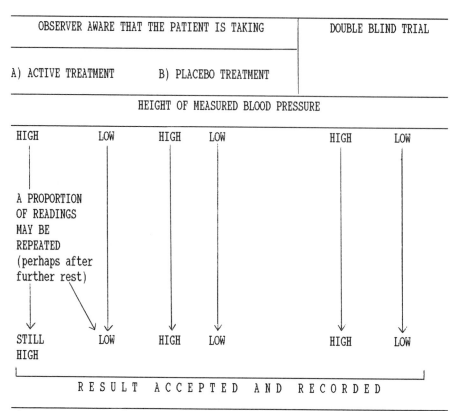

Figure 7-2. Hypothetical blood-pressure readings in an open and a double-blind trial.

Ederer [140] discussed the discovery of n-rays by the eminent French physicist Blondlot. These rays were said to be given off by metals and to increase the luminosity of cards painted with luminous paint. In addition, the rays, when falling on the eye, increased the person's ability to see objects in a nearly dark room. The existence of these rays was disproved by a double-blind study by Pozdena [166] and also by a trick in which the American physicist, Wood, substituted a wooden ruler for a metal file and asked Blondlot whether the file was producing n-rays. Blondlot assured Wood that the file enabled him to see much better [167]. Pozdena performed a straightforward but elegant double-blind trial in which an assistant opened and closed a shutter, releasing and interrupting the flow of hypothetical n-rays. Pozdena recorded increased luminosity as often when the shutter was open as when it was closed [166].

Blondlot's error arose from the difficulty of making subjective assessments. Observer bias can present great problems when determining subjective impressions such as the presence of symptom side effects; this is discussed in chapter 15. More objective measures may also be influenced by observer bias; chapter 13 discusses the repeatability of measurements and the quality control of data.

The measurement of blood pressure in a trial of an antihypertensive drug may be taken to illustrate the difficulties. Let us assume that the protocol specifies that the blood pressure has to be taken after 5 minutes rest in the lying position. Figure 7-2 illustrates the possible sequences of events, first when the observer is aware of the treatment being prescribed and second when the trial is double-blind and the observer is not aware of the treatment being given. For each instance the hypothetical result of detecting a high or low reading is charted. It is possible that all first readings are accepted and recorded except for high readings when the patient is known to be receiving active treatment. In this case the observer may assume, possibly rightly, that the patient has not relaxed sufficiently or that the blood pressure has not been taken correctly. In this case the cuff may be applied more carefully (for example, with the inflatable section more accurately over the brachial artery); the patient may be asked to relax for a further period, and the readings repeated. We assume that the second reading is accepted and recorded. The observer may be correct when he substitutes the second lower reading for the initial high reading. However, high measurements will not be repeated when the patient is known to be taking a placebo and the blood pressure recordings, on average, will be biased towards lower readings on the active treatment. No such bias is to be expected in a double-blind trial.

Fletcher [168] stated, *"Both in initial assessment of the patients and the subsequent assessment of*

*their progress the tests should be applied by observers who
remain unaware of which patient is undergoing treatment and
which is a control. If this is not done, the subjective
judgements which are inseparable from nearly all tests in
clinical medicine may prejudice the results..."*
Another example where objective measurements
are subject to observer bias was given by Kahn
and his colleague [169] when reporting on
measurements of serum cholesterol. They found
that when technicians were given blind duplicate
blood to measure, the standard deviation of the
duplicates was 2.5 times as large as when they
were given labelled duplicates. Wilson concluded,
*"No human being is ever approximately free from these
subjective influences; the honest and enlightened
investigator devises the experiment so that his own
prejudices cannot influence the result. Only the naive or
dishonest claim that their own objectivity is a sufficient
safeguard..."* [170]. Observer bias can favour or
prejudice a positive result.

Observer bias favouring a positive result

Muench's Second Law states: *"Results can always be
improved by omitting controls"* [171]. A control group
can demonstrate the occurrence of spontaneous
improvement and thereby reduce the magnitude of
any overestimated treatment effect. Even with
controls, the unblinded observer may report more
favourably on the new treatments under
investigation. When the trial is double-blind,
observer bias may be prevented and produce a more
correct estimate of the effect of treatment.
Foulds reviewed studies of antidepressant
drugs conducted between the years 1951 and 1956
[172]. He identified 36 studies in the American
literature and 36 in British journals. Only four
trials in the American literature included
controls. In the British literature 16 trials
included controls and, on average, these papers
reported a 19 percent success from treatment. In
contrast, the 20 uncontrolled studies reported an
85 percent success rate. Foulds also made the
interesting comment that patients with anxiety
states tended to improve spontaneously, thus when
subgroups were analysed these patients appeared

to have responded best to treatment. Without adequate controls, it might be erroneously concluded that the antidepressant drugs were most effective in patients with anxiety states.

Observer bias prejudicing a positive result

This theoretical consideration has been discussed by Ederer [140], who suggested that a bias may occur against the experimental treatment *"when the investigator's bias is against the treatment or when he overcompensates for his known bias in favour of treatment. Thurber's moral, 'you might as well fall flat on your face as lean over far backward' (is apt)."*

BIAS ARISING FROM THE ANALYSIS

There are many ways in which the analysis of trial results could bias the conclusions. An example was discussed previously. In the trial of anti-smoking advice [164] the results in the intervention group could be improved by omitting those who did not stop smoking. The intervention, however, was the giving of anti-smoking advice and the total intervention group has to be compared with the total control group as the trial was not intended to be an explanatory study. Similarly, the Anturane Reinfarction Trial [65] has been criticised for leaving out patients in the intervention group simply because they died while the serum concentration of Anturane (sulfinpyrazone) was thought to be negligible (see chapter 19). It has therefore been suggested that trial results should first be examined without the analyst being aware of the treatment given to the various groups, the treble blindness defined above.

The greatest problem in analysis arises from the difficulty in deciding whom to include in the analysis and whom to omit. Similarly during the course of the trial withdrawal of patients may bias the results.

BIAS DUE TO THE WITHDRAWAL OF PATIENTS FROM A TRIAL

Bias may result from withdrawal from the group as a whole prior to randomisation (for example, during a placebo run-in period) or by selective withdrawal from the different treatment groups after randomisation.

Withdrawal of patients prior to randomisation

Many subjects are considered for a trial who are not entered into the study. The subjects may not reach the eligibility requirements or may refuse consent. In addition, a placebo run-in period may be part of the trial and compliance with therapy estimated during this period. If noncompliant patients are withdrawn prior to randomisation only compliant subjects will be entered into the trial and the conclusions from the trial results are only valid for such patients. The problem is discussed in chapter 13 and current practice is to keep a log book containing details of all subjects considered for a trial, irrespective of their subsequent entry to the trial.

Selective withdrawal of randomised patients from a control group

Control treatment such as no treatment or placebo therapy may be less effective than active treatment and patients may be withdrawn owing to treatment failure. In long-term trials of antihypertensive agents to determine their effect on mortality, more patients are withdrawn from the placebo group than from the actively treated group owing to marked rises in blood pressure. If a withdrawal criterion states that a patient must be withdrawn with a certain elevation in blood pressure then the assumption is that, left in the trial, this patient would be likely to suffer an end point such as a stroke. If the withdrawn patients are omitted from the analyses, the placebo and actively treated groups will no longer be comparable for important characteristics, as fewer patients with increasing blood pressure will be removed from the actively treated group [121]. When withdrawn patients are included in the analysis as in the standard intention-to-treat analysis their survival is taken to be representative of all subjects randomised to their particular treatment group (chapter 14). However, if a placebo treated patient is withdrawn and actively treated (and active treatment does confer a benefit) then his or her survival will not be typical of the control group. The alternative is to assume that they suffered an adverse event with a defined probability. An estimate may be available of the probability from observational studies or from earlier trials when patients with the given level of risk were not withdrawn but continued in the study. It may then be possible to allocate some estimate, say, three cardiovascular end points for every ten such patients withdrawn from the trial. To my knowledge, such a correction has not yet been attempted in a trial analysis and therefore the benefits or otherwise of such a manoeuvre have not been established.

Selective withdrawal of patients from the actively treated group

In a double-blind trial, an excess of withdrawals from the actively treated group compared to the placebo group may be expected to represent the frequency with which the active treatment produces adverse effects. However, in a trial that is not double-blind patients with possible adverse drug effects will only be withdrawn from the actively treated group.

In a single-blind trial a patient may suffer an episode of illness that he attributes to the drug he is receiving. Figure 7-3 illustrates the sequence of events that may occur. The patient may have a gastrointestinal upset, rash, influenza, or other condition unrelated to treatment. If he has just started therapy, he may attribute the symptom to the drug or other treatment and not to an intercurrent illness. When the patient is on active treatment, the doctor may agree with the diagnosis and withdraw the patient from the trial with a possible adverse event. When on placebo treatment, the

144

SINGLE-BLIND TRIAL

PATIENT ON PLACEBO   PATIENT ON ACTIVE TREATMENT

GASTRO-INTESTINAL UPSET SHORTLY AFTER STARTING TREATMENT AND UNRELATED TO
TREATMENT.

PATIENT ADVISED TO   PATIENT WITHDRAWN WITH A
CONTINUE WITH PLACEBO   POSSIBLE ADVERSE REACTION

Figure 7-3. Mechanism whereby the adverse effects of treatment may be
overestimated in a single-blind trial.

patient will not be withdrawn and the losses from
active treatment will be greatly in excess of the
withdrawals from placebo treatment. This
situation is avoided in double-blind trials.

It must be remembered that placebos do
occasionally produce an adverse effect. If a
placebo tablet contains lactose it is possible
that diarrhoea will be produced in susceptible
individuals and in this event any excess
incidence of diarrhoea with the active treatment
may be underestimated rather than overestimated.
Usually however, a placebo group and double-
blinding gives an unbiased estimate of the
frequency of adverse drug reactions (see chapter
16).

BIAS DUE TO FAULTY METHODOLOGY

When the trial protocol is not identical for both
the treatment under investigation and the control
treatment, then biased results may be obtained.
Two examples may be provided: (1) when the
treatment has an additional but nonspecific
effect this may be difficult to reproduce in the
control group, and (2) when the protocols for the
two groups differ in order to save time or money.

The treatment may have a nonspecific effect that is difficult to reproduce in the control group

The nonspecific improvement resulting from giving drug treatment can be estimated from a placebo group in a single or double-blind trial of drug treatment. The nonspecific effect cannot be estimated when surgery is the intervention strategy and may also be difficult to assess when therapeutic advice is given. In a trial of antismoking or dietary advice, the patients in the intervention group may be seen repeatedly and it will be difficult to provide placebo counselling on any neutral topic. First, it would not be ethical to advise the control group against an activity that does not harm them. For example, in a controlled trial of lowsalt dietary advice in hypertensive patients poorly controlled on antihypertensive drugs, we decided that only the patients randomised to the diet would get dietary counselling. However, interviews were arranged for both the intervention and control groups to assess compliance with drug medication and to assess the patients' quality of life [173]. The prohibition of a neutral dietary constituent as control advice might have been acceptable in the short term (for example, ice cream could have been prohibited). However, such advice would not have carried much conviction or a reduction in ice cream consumption may have affected blood pressure control. To give the control group major and therefore comparable advice (such as low-cholesterol diet) would not have provided an untreated control group. There may be no very satisfactory method of performing a double-blind trial of dietary advice but the effect of a low salt intake can be determined by advising all patients to take the low salt diet and add either salt tablets or placebo tablets according to whether they were in the control or intervention group.

The protocol differs between the intervention and control groups in order to save time or money

A theoretical example may be employed to illustrate a faulty protocol that requires the intervention and control groups to be treated separately. A new drug has to be tested against placebo treatment and blood tests and blood measurements are required for the active treatment group, but only blood pressure measurements in the placebo group as a considerable amount of data have shown that placebo treatment does not affect the blood tests. However, taking blood in the actively treated group may raise blood pressure in that group either because the blood test precedes taking the blood pressure or because the patient anticipates the blood test. No such rise will be observed in the group given placebo. Whenever possible the protocol for the actively treated and control groups must be identical.

THE USE OF THE DOUBLE-BLINDING TECHNIQUE

The double-blinding or double-masking technique is very important and should be used, whenever possible in randomised controlled trials. Knowelden [174] has commented,

When everyone is in the dark, subjective measures can be used with confidence as there can be no bias introduced by patient or observer. If an observer had to diagnose a slight paroxysmal cough in a child known to have been vaccinated against whooping cough he might, because of bias in favour of the vaccine, decide it could not be pertussis. What is probably more likely than such cheating is that the clinician would attempt to compensate for his known bias and label the case one of whooping cough against his better judgement.

Double-blindness, however desirable, is not always possible and many trials have been performed without any blinding or with only the patients being unaware of the treatment (single-blinding). We must discuss how double-blinding can be achieved, when it cannot be utilised, why it is often not attempted, and what may happen if it fails.

How to achieve double-blinding

The following steps will lead to double-blinding:
1. The patient must understand that the trial is double-blind and give written informed consent.
2. The control treatment and experimental treatment must be identical. If tablets are used, they must be the same size, shape, colour, taste, and smell.
3. Details of the exact treatment being given must be held in a secure position and be immediately accessible in an emergency. The treatment is often written on a card and enclosed in an opaque, sealed envelope. When the treatment is a drug these envelopes may be conveniently held in the pharmacy responsible for dispensing the treatment.

When masking is not desirable

Blindness is not desirable when it would result in unacceptable suffering or discomfort for the patient. The taking of dummy tablets will not be expected to produce any discomfort but sham operative treatment or dummy treatment involving repeated injections would do so. In the early Medical Research Council trial of antituberculosis therapy [5], it would not have been desirable to inflict dummy injections on the control group for several months. Similarly, although placebo operations have been performed (chapter 3) it is only possible to perform sham operations in humans in very exceptional circumstances. These reservations have not applied to animal experiments where sham operations are the rule.

When masking is difficult

Double-blinding techniques may be difficult when the treatment has obvious effects or when the treatment involves a change in life-style.

i)   The treatment has an obvious effect

Certain drugs have specific actions which, if detected, would enable the patient or observer to break the treatment code. Examples are given by the beta-adrenoceptor blocking drugs that lower the pulse rate, and diuretics that usually result in an increase in serum uric acid. An observer could use this information to break the code and steps should be taken to ensure that he does not become aware of these results.

ii)   The treatment involves a change in life-style

Dietary change is an example of an alteration in life-style where masking is difficult. Unlike trials of drug treatment the advantages of double-blind trials of dietary advice are balanced by a number of disadvantages. In a double-blind trial, items of diet have to be provided by the investigator, some of which contain the dietary constituent under investigation and are given to one intervention group and some that appear identical but do not contain the constituent and are given to a second group. An example is given by the National Diet Heart Study [175] in which saturated fats were the dietary constituent under investigation.
    Table 7-7 lists the advantages of double-blind and non-blind trials of dietary advice, assuming that the trial end points are not altered by lack of blindness, being either easily ascertained (for example, death) or determined in a blind manner (for example, measurements of serum cholesterol). A non-blind trial of dietary advice tests the combined effect of diet and increased medical attention. Those who receive dietary advice must have greater contact with the research workers and this may affect the subjects' well-being in a nonspecific manner. The subject is aware that he has altered his diet and an increased attention to his food may either improve or reduce his sense of well-being. In a double-blind trial the intervention and control group will be equally affected by nonspecific dietary changes whereas in a nonblind trial the control group may have an unrestricted diet.

Table 7-7. Advantages of double-blind and non-blind trials of dietary advice. With a double-blind trial the subjects are given certain items of food whose composition is unknown.

Double-blind dietary study	Non-blind dietary study
1. Allows effect of diet to be assessed over and above nonspecific effect of dietary advice.	1. Assesses combined effect of dietary advice and taking the diet.
2. Changes in diet in control group less likely to occur (less contamination of control group).	2. Control group has an unrestricted diet.
3. Easier to secure unbiased ascertainment of all end points.	3. Easier to administer the trial and much less expensive.
4. Easier to secure motivation of participants in control group.	4. Fewer ethical problems (patients and physicians know the treatment).
5. Withdrawal rates should be similar in different groups.	

However, in a non-blind trial some of the control group may alter their diet in a similar manner to that of the intervention group. If the control group hears about the intervention and alters its diet, the group is said to be contaminated. Double-blindness should ensure that the trial end points are determined similarly in the different groups and that the drop-out rates are not affected by knowledge of the treatment. These advantages are balanced by a great increase in complexity and expense, possible ethical problems, and when monitoring adherence to the diet by closely monitoring blood or urine tests, these results must be withheld from the investigator. Monitoring of blood or urine may be used to improve compliance but this strategy can be employed in a double-blind or an open trial.

Strategies to be employed when double-blindness is not possible

When a treatment cannot be provided blind or when it produces obvious effects it may not be possible for one investigator to conduct a masked trial on his or her own. If the effects of treatment are determined from laboratory investigations the trial can be designed so that laboratory results are withheld from the investigator. When a clinical measurement would detect the treatment, one investigator can determine the result and a second can be responsible for the care of the patients. Measurements may also be limited to those that are not open to observer bias (hard end points).

i) Blind the investigator to certain laboratory tests

An example of this procedure may be provided by considering a trial of the blood pressure lowering effect of tienilic acid, a drug that markedly reduces the serum uric acid. During a double-blind trial the investigator measured blood pressure and interviewed the patients but the results of serum uric acid measurements were withheld from him [91]. It was planned that if a serum uric acid result was grossly abnormal and action required, the investigator should be told and the treatment code broken.

ii) Two investigators: one to assess the patient and the other to provide treatment

Two investigators may have to be involved when therapy cannot be provided blind or when a treatment affects a measurement in a constant manner. One investigator provides the treatment or takes the measurement and the second assesses the end points of the trial. An example is provided by a trial of dietary advice where one investigator gives the therapeutic advice and a second assesses the outcome in terms of weight or blood pressure without knowing the treatment allocation. Similarly, in a trial of medical treatments for peptic ulceration, it may be possible for a second observer to assess the results of the treatment from radiological plates or photographs of the ulcer.

In certain circumstances, the person providing the treatment may be masked and the person taking the measurements may not. Such an arrangement would be appropriate in a trial of the antihypertensive effect of increasing doses of a beta-adrenoceptor blocking drug compared with a different antihypertensive drug. Beta-blocking drugs lower both pulse rate and blood pressure and if the investigator knows the pulse rate, blindness would be lost. One investigator should measure blood pressure and pulse rate. The second investigator should remain blind and be given measurements of blood pressure but not pulse rate unless this is abnormally low. The second person would then be responsible for prescribing increased doses of the drugs and assessing any side effects of treatment.

iii) Hard end points

When masking is not possible, the analysis may be confined to so-called hard measurements. A hard end point is one that is not open to biased ascertainment; the best example is the fact of death. When survival is to be determined blindness is not necessary. However, although the fact of death may be incontrovertible the cause of death may be open to doubt. For example, sudden death may or may not be due to myocardial infarction and it may be difficult to distinguish cardiac from respiratory death. The difficulty is compounded when the investigator only knows the cause of death as written on the death certificate. These certified causes are usually not the result of postmortem examinations and are subject to error. Moreover, if an observer believes that a particular treatment prevents, or causes, a specific cause of death, this preconception may lead to a biased assessment of the cause of death.

The disadvantages of blinding

It is possible that masking will adversely affect patient recruitment and investigator participation. Identical control treatment will have to be provided at extra expense and it may be difficult to ascertain the exact treatment in an emergency. With drug treatment, it will be necessary for the patients to take extra tablets or capsules and labelling errors may occur. Lastly, masking may require additional personnel.

i)   Fall in patient recruitment

When patients are informed that they and the investigator are unaware of the treatment they are to receive, some may be dissuaded from taking part in the trial. Similarly, they may dislike the idea of taking placebo tablets. In my experience this has only rarely been a problem.

ii)   Failure of investigators to take part in the trial

Potential investigators in multicentre trials may be unwilling to take part if a protocol is double-blind, although there are no data to substantiate this claim.

iii) Provision of identical control treatment

Making tablets that are identical in appearance, touch, and taste may be difficult. The problem may be simplified by making identical capsules to contain the active drug or the placebo. However, if an active pharmaceutical compound is dispensed in a capsule, this formulation will differ from that of any tablets normally available and may affect the bioavailability of the drug.

   If a control tablet is manufactured of an identical size, colour, and shape it may still smell and taste differently or differ in some other way from the experimental treatment. There is a story of a patient who reported that he knew his tablets had been changed at the last visit to the clinic as the new tablets were difficult to flush down the toilet! It is hoped that most tablets will not be disposed of in this or any similar manner.

   It may prove possible to make an acceptable placebo for an active treatment but not to make two active treatments identical. In this instance, the patients can be asked to take two sets of tablets throughout the trial: one representing treatment A (active or placebo) and one treatment B (active or placebo). This procedure has been called the double-dummy technique and is especially useful in a cross-over trial. In the first phase of the trial a patient takes one active drug and the placebo corresponding to the second drug, and in the second phase the active second drug is taken together with a placebo of the first drug.

iv) The patients have to take more tablets or capsules

In a drug trial, taking a placebo always increases the number of tablets or capsules to be consumed. This is acceptable in order to make an unbiased comparison of an active treatment with an inactive treatment, rather than a test of an active treatment against no treatment at all. However, extra tablets may be required to preserve blindness, as with the double-dummy technique discussed above.

v) Expense

When identical tablets or capsules are provided, it is expensive to manufacture, label, and distribute them. Dispensing may also be more complex in a double-blind drug trial.

vi) Errors in labelling

In the labelling and distribution of identical tablets, great care must be exercised. The containers are only identified by code numbers or letters and the labels have to be checked and all stages carefully documented. I know of two trials where errors of labelling have occurred so that active treatment has been given instead of placebo and vice versa. Fortunately the errors were limited to only one or two patients. In one trial the errors were discovered when patients were checked for compliance: a patient on placebo had active drug metabolites in his serum when he had had no access to any active drug. In the second trial a patient on large doses of placebo suddenly received large doses of an active antihypertensive drug. The resulting side effects revealed the error.

vii) Difficulties in breaking the treatment code when necessary

As discussed earlier, the code must be available for consultation in an emergency; it should be held in sealed envelopes, one envelope for each patient, and accessible 24 hours a day. The codes can only be held centrally if a coordinating office is manned day and night and otherwise must be held at a convenient place (for example, in the hospital pharmacy). The subjects in the trial should be given details of whom to contact in an emergency and these persons must know the location of the codes.

viii) The provision of extra personnel

The need for two investigators to ensure double-blindness (see above) may be both difficult and expensive.

The consequences of a breakdown in double-blindness

The breakdown of double-blindness may affect the observations made in a trial and have serious consequences. In one trial of vitamin C against placebo the active compound reduced the frequency and severity of the common cold [176]. However, this result was contrary to the results in other trials and the subjects were asked which tablets they thought they were taking. The subjects guessed correctly more frequently than they should have done by chance alone. This was a worrying finding, as those who knew they were on placebo may have reported more colds and those who realised they were on active treatment, less colds. Alternatively, a positive answer could have been masked if those on vitamin C, as a consequence of this knowledge, had not eaten as much fresh fruit and vegetables as those who thought they were on placebo. As the trial was to determine whether or not the vitamin had a non placebo effect on the incidence of colds, the wrong answer may have been obtained.

It is good practice to ask each patient about the treatment they guess they are taking.

In the National Diet-Heart Study [175], participants were asked to purchase special dietary foods with different fat contents and the trial was conducted in a double-blind manner. The participants were asked about the amount of fats in their diets and 43 percent in each dietary group considered that their diet included a large reduction in total fat. There had been no loss of double-blindness and dropout rates were independent of diet as were the proportions volunteering for further study. Similarly, the doctors and nutritionists involved in the trial were asked to specify the diet that had been assigned to the individual patients. The correspondence between the actual diets and their guesses was no better than would be expected by chance.

A consequence of the breakdown of blindness has been examined in a trial where the investigators were apparently able to identify the active treatment. Heaton-Ward [177] performed a double-blind trial to assess the effect of a monoamine oxidase inhibitor (Niamid) on the activity and behaviour of subjects with Down's syndrome. The observers were told that the trial was of a cross-over design but at the time of cross-over the same treatment was continued. The observers reported an initial improvement in the actively treated group but not in the placebo-treated group. However, after the supposed cross-over, they reported a deterioration in those who had initially improved and an improvement in those they first imagined had not improved. The objectivity of the observations appeared to be in some doubt. Abraham et al [178] termed this bias and its result the *Heaton-Ward effect*, after the author of this trial. The trial was concerned with activity and behaviour and in this field subjective impressions bedevil the interpretation of results (see chapter 15). A Medical Research Council Trial into the effects of antidepressant drugs [179] showed that a breakdown of blindness due to side effects led to differences between treatments being observed in subjective assessments. The effects of treatment were not confirmed by a more objective reduction in the length of stay in hospital.

Double-blinding is very important, and Beecher commented, *"... there is evidence in surgery as in other fields, that the enthusiast actually gets results which are better than those of the sceptic"* [55]. In a double-blind trial both the subjects and investigators should be interviewed to determine whether or not masking has been preserved.

THE USE OF PLACEBOS

When an effective treatment is available this is usually employed in the control group. However, in the short term, or when no effective treatment has been discovered, placebos should be employed as control treatment whenever possible.

Why should placebos be used?

Bradford Hill stated, when discussing the clinical trial, *"To some patients a specific drug is given, to others it is not. The progress and prognosis of these patients are then compared. But in making this comparison in relation to the treatment the fundamental assumption is made - and must be made - that the two groups are equivalent in all respects relevant to their progress, except for the difference in treatment"* [1]. As discussed above, the use of placebo treatment in the control group ensures that any difference between the actively treated and the control group is due to the active constituent employed in the trial and is not a nonspecific effect of giving any treatment.

When should placebos be used?

Placebos should *not* be used as control treatment when there is a definite evidence that withholding available treatment may be detrimental to the patient's health. Other reasons have been advanced for not employing placebo treatment: the increase in cost of the trial due to placebo materials, distribution, coding, documentation, and the increase in the work load of the investigator. However, when there is no treatment of proven worth, a placebo-controlled trial is to be preferred to one using an untreated control group. In some trials the use of a placebo is obligatory; in other trials it is advisable but optional.

i)   Essential uses of placebo

Placebo treatment is essential in trials of anxiolytic, hypnotic, and anti-inflammatory drugs. The use of a placebo regime allows the day-to-day variation of subjective sensations such as pain to be measured together with any spontaneous improvement with time or as a nonspecific response to tablets.

ii)   Important uses of placebo

Placebos are often employed in chronic conditions such as hypertension and diabetes mellitus when available treatment is known to correct the pathophysiological abnormality but has not been shown to reduce mortality or morbidity. Three examples where active treatment has not been proven to be of benefit to the patient's health are mild hypertension (diastolic pressure 90-95 mmHg), benign hypertension in the very elderly, and mature-onset diabetes. The long-term treatment of these conditions with a placebo is justified as no known treatment confers definite benefit in terms of reduction in mortal or morbid events. In hypertension, the use of a placebo is particularly important as the control group may exhibit a reduction in blood pressure owing to the reassurance offered by supposedly active treatment. However, a specific placebo effect in reducing blood pressure has not been proven and the reduction in pressure may reflect the process of familiarisation with medical attendants and clinical surroundings.

iii) Other uses of placebos

In the chronic diseases discussed previously, placebo treatment is often employed in the short term to determine a baseline level of blood pressure or blood sugar even when the patient is known to require active treatment in the long term. A frequent example of the use of a placebo is in patients with moderate or severe hypertension. The investigators know that active treatment must be given eventually, but the treatments under trial are to be compared with a baseline untreated period. Taking a placebo ensures that the control period is more similar to the periods of active treatment and that the subjects have a sustained level of blood pressure despite placebo treatment. Two strategies partially resolve the ethical problems: the length of the placebo treatment is limited to a short period of 3-4 weeks, and the placebo treatment is stopped if blood measurements exceed an arbitrarily defined level.

How placebos should be employed

i)   The agreement of the subject must be obtained

The patient must agree to take either the placebo or the active treatment. Fortunately, many patients are prepared to serve as experimental subjects and contribute to the common good. Patients presenting to a doctor are usually willing to adopt the advice they are given and if the doctor suggests taking part in a placebo-controlled clinical trial, the patients tend to accede to the request. They may be unwilling to make an independent choice between entering or not entering the trial. Such indecision might be analogous to the situation where an airline pilot, with a faulty aeroplane, asks the passengers whether he should go on or turn back. The doctor's advice will be taken and the trust of the subjects must not be abused.

All relevant information about the trial must be provided for the subject, including the following:

1.   The fact that a placebo is being used in the trial and the patient may receive it.
2.   Whether or not the trial tests a concept that could lead to a benefit for the patient.
3.   That any new treatment has been adequately tested in the laboratory and may be an advance over existing therapy.

4. That the subject may withdraw from the trial at any time without affecting subsequent medical treatment and without giving a reason.

The patient should be given a written copy of this information to facilitate full comprehension and be asked to provide written consent to taking part in the trial. Such consent does not reduce the investigator's responsibility but does provide proof that the patient read about the trial and agreed to take part on the basis of the information. It also proves that the investigator discussed the trial with the patient (chapter 3)

Bradford Hill questioned whether the patients should be told that they may receive a placebo and he wrote, *"Having made up your mind that you are not in any way subjecting either patient to a recognised and unjustifiable danger, pain or discomfort, can anything be gained ethically by endeavouring to explain to them your own state of ignorance and to describe the attempts you are making to remove it ... Once you have decided that either treatment for all you know may be equally well exhibited to the patient's benefit, and without detriment, is there any real basis for seeking consent or refusal?"* [36]. Many would support this view for a trial of acute treatment in a stressful situation (for example, on admission to a coronary care unit with acute myocardial infarction). Under less stressful conditions and with long-term treatment informed consent should be obtained.

ii) Single or double-blind?

The trial employing a placebo must be single or double-blind and the placebo treatment must be identical to the active treatment.

iii) Randomise the order of placebo treatment in a cross-over trial

A placebo run-in period is often employed in randomised trials to ensure that only patients with certain characteristics enter the trial. An initial period on placebo treatment allows patients to be excluded if they do not have, for example, a persistent elevation in blood pressure, blood sugar, or serum cholesterol. Subjects who do not comply with the trial protocol may also be identified during this period. However, when a period of placebo treatment is to be compared with an interval on active treatment in the same patient, the order of treatment must be varied, using random allocation. In some trials the placebo treatment has been given first and followed by the randomisation of two treatments. When the investigators are comparing the two treatments with placebo, the effect of the drugs in lowering blood pressure may be exaggerated, as the average blood pressure tends to fall throughout treatment. It is not acceptable to report a placebo-controlled randomised double-blind cross-over trial of two treatments when the placebo period always precedes the randomised part of the trial. In a trial of two antihypertensive drugs the reader would assume that the baseline blood pressure was accurately determined by a randomised period of placebo treatment during the trial. Baseline measurements should be obtained double-blind during the trial with the two active treatments and the placebo treatment given first, second, and third with equal frequency.

Disadvantages of the use of placebos

The disadvantages arising from the use of placebos overlap those given for double-masking. In addition, the use of a placebo may disrupt the doctor-patient relationship and produce medico-legal problems.

i)   Disruption of the doctor-patient relationship

Bradford Hill stated, *"The doctor will also wish to consider the doctor/patient relationship. Harm may be done if the public comes to believe that doctors are constantly using them as guinea-pigs. In exhibiting new treatments they are, it is my belief, doing that willy-nilly, but the public does not realize it. But they need not go out of their way to make it obvious by an unnecessary use of dummy pills"* [36]. But what is an unnecessary use? The provision of a comparable control group by the use of placebos is often essential.

ii) Medico-legal problems

It is possible to argue that if a drug has been shown to be beneficial in one clinical trial against placebo, an investigator repeating the trial is knowingly placing at risk any patient treated with placebo. If patients in the placebo group suffer harm, the investigator may be sued for damages on the grounds of negligence. Dollery [155] stated, *"Lawyers appear to have little time for the contention that there is uncertainty about the efficacy when only one or perhaps two trials have completed. They do not understand the concept of statistical uncertainty and are accustomed to resolving doubts about factual uncertainty in the courts."*

An investigator may also run into problems when he employs an experimental treatment and there is no definite evidence that it will be successful. If the experimental treatment subsequently proves of benefit in later trials, the courts may examine the original trial and find in favour of subjects in the placebo group who have suffered harm. Dollery provides the example of trials of active antihypertensive treatment in patients with an untreated diastolic blood pressure of 90-105 mmHg. This group has only mild hypertension and the original trial, the Veterans Administration Co-operative Study on Antihypertensive agents [52], found a 35 percent reduction in complications in this group with active treatment; but this did not achieve statistical significance. Moreover, the trial included a high proportion of patients who had already suffered a complication of hypertension and patients were only included if the high blood pressure was maintained during a hospital admission. For these and other reasons further trials of antihypertensive therapy have been performed for mild diastolic hypertension in young and middle aged patients. Two trials were soon completed and shown a benefit in patients with a diastolic pressure over 100 mmHg but not 90-99 mmHg [180-181]. As the trials subsequently found a benefit from active treatment when the initial diastolic pressures were between 100 and 104 mmHg, can such patients take legal action when they have sustained a disabling stroke while on placebo in the original or later trials?

If the courts find in favour of the patient, will they realise that they do this with the assistance of hindsight? When the first trial was started, for all the investigators knew the patients treated with placebo may have been the fortunate ones. However, many would claim that patients who suffer from taking part in randomised trials should be compensated. In this event the compensation payable should not be excessive as pharmaceutical companies and other organisations may be unwilling to support placebo-controlled trials. Imagine the problem if trials of new treatments are conducted only against a supposedly active treatment, the active treatment being of no use. There could be a proliferation of useless drugs all being as good as the active (useless) treatment.

How many positive trials are required against placebos?

How many trials of active versus placebo treatment have to show a positive result before the results can be accepted and further use of placebo prohibited? In chapter 19 the problem is discussed of how many trials of a particular treatment are required before we are able to reach a confident conclusion. Only two trials of antihypertensive treatment were performed before it was accepted that middle-aged men with a sustained diastolic pressure over 105 mmHg should be treated and not exposed to long periods of placebo treatment. However, when trial results are contradictory, as with the use of anticoagulants following myocardial infarction, a large number of trials may have to be performed. The International Anticoagulant Review Group examined nine trials where proper control groups were employed and concluded that the benefits from anticoagulants were of the order of a 20 percent reduction in mortality in men and zero in women [76]. The medical profession has not considered this to be a worthwhile gain in men and the treatment has not been employed in most countries (see chapter 19).

The recent publication of the results of the Systolic Hypertension in the Elderly Programme (SHEP) trial [182] caused much debate for those involved in the Systolic Hypertension in Europe (SYST-EUR) trial which was in progress [183]. Both trials compared active treatment for isolated systolic hypertension (ISH) with placebo treatment, and the SHEP trial reported a benefit in terms of a reduction in non fatal stroke and cardiac events. Should the SYST-EUR trial continue or indeed should new trials of treatment be allowed to start?

The trial continued for the following reasons:

1. The benefit : risk ratio was not unequivocally in favour of treatment in the SHEP trial as the adverse effects of treatment were not trivial and mortality was little reduced [184,185].

2. The generalisability of the SHEP results is not known and needs to be established [186]. Similar concerns about generalisability have led to the Swedish Two County Study of mammographic screening (which lowered mortality from breast cancer) being repeated in the UK [187,188].

3. A Bayesian statistical approach suggested that in the SYST-EUR trial there is a 64% chance of not finding a minimum reduction in stroke of 15% at P <0.01 [189]. There was therefore no guarantee that the SYST-EUR trial will find a similar result to the SHEP trial.

A positive result may occur by chance in up to 5 percent of trials (according to the level of significance required in the trials) and it would appear reasonable to require two positive trials before accepting any results as conclusive. A second trial, including placebo treatment, may usually be started in order to confirm or refute the initial finding. It is arguable, however, that very large trials should not be repeated. The first trial should act as a provisional guide to clinical practice while further evidence is awaited.

CONCLUSIONS

The results of a trial may be expected to be free of bias when randomisation has produced an equivalent control group. The control group must provide an estimate of any placebo effect, an estimate of any time trend in the condition being investigated and of the effect of regression to the mean.

Noncompliance with therapeutic advice may alter the response to treatment but patient and observer bias may be removed by single- and double-blinding respectively.

Great care must be taken in the analysis of the results of the trial in order to avoid introducing bias, especially when considering subjects who withdraw from the trial.

The method of conducting the trial is very important and double-masking and placebo control should be introduced where possible. Both these techniques can present difficulties in the form of complexity, expense, and even litigation. If double-blinding, placebo control, and randomisation are not employed the consequences are likely to be more serious than any adverse consequence from using them.

8.   WRITING THE PROTOCOL

The protocol, or as Bearman preferred, the manual of operations [190], may have to serve many functions: raising monies from a funding agency; obtaining the approval of an ethical committee; recruiting participants; providing a detailed and specific list of instructions on how to perform the trial; and lastly supplying a permanent record of what was intended in the trial. The same document may serve all these functions, and in addition a section on finances may be included when the document is used as a grant application.

The protocol should consist of clear statements on the following: where the trial is being run and by whom; the background of the trial; objectives; numbers to be entered; eligible patients or subjects; procedures to be adopted during the trial; duration over which the trial will be performed; handling of dropouts; proposed analyses; criteria for stopping the trial; publication policy; and financial considerations. A copy of all the documents to be used in the trial should be attached as well.

WHERE WILL THE TRIAL BE PERFORMED AND BY WHOM?

It may appear self-evident that the personnel involved and the site of the trial should be stated. However, in the case of long-term, often multicentre trials, many difficulties may arise. McFate Smith [191] discussed an organisational model for a multicentre trial and considered that failure to specify the responsibility of various committees has led to difficulties in the execution, analysis, and publication of large trials. The persons involved in running the trial must be stated and can be grouped as follows.

The steering committee

The steering committee will include the principal investigators and a chairman should be named. This committee should meet with a predetermined frequency and should control the running of the trial. In some trials, an executive committee is responsible for the day-to-day running and funding of the trial, and a steering committee for the protocol [192].

The coordinating centre

The coordinating centre will include staff with clearly identified responsibilities: staff responsible for quality control, data processing, analysis of results, preparation of reports, and distribution of any medications that may be required.

The clinical centres

The staff involved at the clinical centres must be included in the protocol and acknowledged in any publications.

Advisory board or ethics committee

Most large trials have a panel of experts who constitute an advisory board and whose advice may be sought when necessary. The board or committee should be independent of the working members of the trial and the other committees. This committee should decide when the trial should be stopped or recruitment halted [192].

Other committees

Other committees include a monitoring and statistical committee, responsible for performing interim and final analyses and quality control during the trial; an endpoint committee to provide an objective blinded evaluation of previously defined endpoints; and a publications committee to decide on how, when and where the results should be published [183,192].

Funding agency

The source of funds should be stated as should the names of persons responsible for administering the grant.

BACKGROUND TO THE TRIAL

An exhaustive review of the literature is not called for in the protocol but the reasons for performing the trial should be stated and a few references cited in support.

OBJECTIVES

The major and minor objectives must be clearly stated and enumerated (chapter 4).

NUMBER OF PATIENTS REQUIRED

The number of patients or subjects to be entered into the trial in order to achieve the objectives must be stated, together with the assumptions made in arriving at this figure.

ELIGIBLE PATIENTS

The method of recruiting patients should be noted and the criteria for inclusion and exclusion stated explicitly. It is also desirable to collect a limited amount of information on those who were considered for the trial but not included and, where possible, details on patients who were available but not considered. The trial participants may then be viewed as a subset of those available and an impression gained of how representative are the trial subjects of the population.

PROCEDURES TO BE ADOPTED DURING THE TRIAL

This section should include the information to be

given to the subjects, the method of obtaining and recording consent, details on any run-in period, how randomisation will be achieved, the treatment schedules to be employed, blindness, data to be collected, and the end points for the trial. Exact methods must be described for any measurements that are to be made (chapter 13). Methods of recruitment, and entry, exclusion, and withdrawal criteria must be stated precisely.

THE DURATION OF THE TRIAL

An attempt should be made to estimate how long it will take to recruit the required number of patients. The financial support necessary will depend on the expected length of the trial.

WITHDRAWALS FROM THE TRIAL

The protocol must state under what conditions patients may be withdrawn from the trial and how they should be followed and treated thereafter. The details will include those adverse reactions that require withdrawal and how such reactions, and lesser side effects not leading to withdrawal, are to be detected and treated. It is important to follow all patients who are withdrawn.

ANALYSIS

The outline proposed for the analysis of the results should be stated and, most importantly, when and how often this analysis will be performed. Details on computer facilities may be required.

CRITERIA FOR STOPPING THE TRIAL

The criteria under which the trial will be terminated should be stated in as much detail as possible. An attempt should be made to foresee all possible eventualities (chapter 12): for example the mortality from one condition in the intervention group may be reduced with statistical significance, yet total mortality may not be reduced.

PLANS FOR ORAL PRESENTATIONS AND PUBLICATION

In a large trial involving many investigators a case may be made for planning oral presentations and publications in advance. It may be very destructive if those who are aware of preliminary results leak this information before all the investigators have been informed and before the final report has been agreed. The protocol may therefore state how the final results should be presented to medical colleagues and to the general public.

FINANCIAL CONSIDERATIONS

Finances should be sought to cover all the costs of the trial. Where possible, funding should be applied for to cover the extra costs of secretarial assistance, computing, stationery, travel, and other overheads. The application for financial assistance should make allowance for these costs and even for monetary inflation if this is expected. However, funding agencies will not usually expect to pay the salaries of the principal investigators or their secretaries; nor will they often be willing to provide office accommodation, word processors, telephones, and other basic items of equipment that would be expected in a standard office.

ADDENDA

The following addenda should be included in the trial protocol.

Documents to be used in the trial

Examples of the recording documents should be attached to the protocol.

Details of methods

Full details of all the methods to be used in the trial must be included in the protocol, usually as addenda. Bearman provided an example of a protocol where it was stated that various biochemical measurements must be within normal limits [190]. The protocol did not state the method of performing the biochemical tests nor the normal limits to be expected. These details must be provided.

Quality control

The protocol must state how the precision and accuracy of various measurements will be determined and followed during the trial. If a trial monitor is to be appointed the protocol should state how this person is to perform his or her duties. Details of the training of staff must also be provided.

SUMMARY OF THE PROTOCOL

For a large trial the protocol will become, of necessity, a bulky and unwieldy document. Each section should be carefully summarised and the protocol be introduced by a brief report of its contents.

CONCLUSIONS

The protocol must be written with great care. An inadequate document, with insufficient information and containing errors, is unlikely to attract financial support and may not meet with the approval of an ethical committee. Also, the protocol cannot be employed as an adequate manual of operations if it lacks the necessary detail. Many large research centres in the United States employ professional writers to finalise protocols, especially when they are used in a grant application.

Appendix 8 is based on the headings in this chapter and allows the investigator or reviewer to check that the protocol has covered most of the important items in the design of a trial. Similar checklists have been prepared by Spriet and Simon [193] and the Clinical Trials Unit of the London Hospital [194]. Undesirable responses to the questions are in italics.

APPENDIX 8: Checklist to assess the protocol for a randomised controlled trial. An undesirable response is in italics.

A1 Who will perform the trial and where?

1.1	Are all the participants named?	Yes ☐ *No* ☐
1.2	Are their addresses given?	Yes ☐ *No* ☐
1.3	Are steering and other committees necessary?	Yes ☐ No ☐
1.4	Is the constitution of these committees stated?	Yes ☐ *No* ☐
		Not required ☐

A2 Background to the trial

2.1	Has ethical committee approval been given?	Yes ☐ *No* ☐
2.2	Are the authors aware of the literature in their field?	Yes ☐ *No* ☐
2.3	Are the authors aware of similar trials that have been completed?	Yes ☐ *No* ☐
	or that are in progress	Yes ☐ No ☐

A3 Objectives

3.1	Are the major objectives clearly stated together with their magnitude?	Yes ☐ *No* ☐
3.2	Are the objectives realistic?	Yes ☐ *No* ☐
3.3	Does the trial answer important questions?	Yes ☐ *No* ☐

A4 Number of subjects required

4.1	Have the numbers been calculated?	Yes ☐ *No* ☐
4.2	Size of type I error ☐ and type II ☐	
4.3	Is the type I error two-tailed?	Yes ☐ *No* ☐
4.4	Can the authors recruit this number of patients?	Yes ☐ *No* ☐

A5 Selection of subjects

5.1	Is previous treatment allowed?	Yes ☐ No ☐
5.2	Are the diagnostic criteria clear?	Yes ☐ *No* ☐
5.3	Is informed consent obtained?	Yes ☐ *No* ☐

5.4 Are selection criteria (age, race, gender) well
thought out?   Yes ☐ *No* ☐

5.5 Do exclusion criteria cover all ethical problems? Yes ☐ *No* ☐

A6 Conduct of the trial

6.1 Is the design: parallel groups? ☐
             cross-over? ☐

6.2 Is the design: open? ☐
             single-blind? ☐
             double-blind? ☐

6.3 If the trial is cross-over, are there data showing
no change in treatment differences with period? Yes ☐ *No* ☐
                    Is there a washout period? Yes ☐ No ☐

6.4 Will there be a pilot trial?   Yes ☐ *No* ☐

6.5 Will blindness be preserved?   Yes ☐ *No* ☐
                                        Not relevant ☐

6.6 Will the observers be adequately trained?   Yes ☐ *No* ☐

6.7 Are the end-point measurements valid
and repeatable?   Yes ☐ *No* ☐

6.8 Are randomisation procedures efficient?   ☐
                    too complicated?   ☐
             or open to manipulation?   ☐

6.9 Is the treatment fixed? ☐
or to be titrated? ☐

6.10 Are the doses reasonable?   Yes ☐ *No* ☐

6.11 Is the labelling and checking of any drug treatment
adequate? ☐
or inadequate? ☐

6.12 Is it clear which accessory treatment will be
allowed?   Yes ☐ *No* ☐

6.13 Will compliance with treatment be determined? Yes ☐ *No* ☐

6.14 Is the subsequent treatment for patients who
complete the trial stated?   Yes ☐ *No* ☐

6.15 Will a trial monitor be appointed?   Yes ☐ *No* ☐

A7 Duration of trial

7.1 Duration for the individual patient   ☐ weeks/months/years

7.2 Duration of recruitment   ☐ weeks/months/years

A8 Withdrawals from the trial

8.1 Will patients who are withdrawn be followed? Yes ☐ *No* ☐

8.2 Will severe adverse reactions be detected
quickly? Yes ☐ *No* ☐
8.3 Will side effects be detected? Yes ☐ *No* ☐

A9 Analyses
9.1 Are the proposed analyses sensible? Yes ☐ *No* ☐
9.2 Will the trial be analysed on the intention-to-treat
principle alone? ☐
the per-protocol principle alone? ☐
or both principles? ☐
9.3 Are any necessary computer facilities available? Yes ☐ *No* ☐

A10 Criteria for stopping the trial
10.1 Are some possible outcomes clearly stated? Yes ☐ *No* ☐
10.2 Have decision rules for stopping the trial been
defined? Yes ☐ *No* ☐
10.3 Will the data be reviewed efficiently as they
accumulate? Yes ☐ *No* ☐

A11 Presentation of the results
11.1 Are regular meetings planned? Yes ☐ No ☐
11.2 Are the plans for presenting the results
acceptable? Yes ☐ *No* ☐

A12 Financial considerations (N/r = not required)
12.1 Are the following staff available when required:
clinicians? N/r ☐ Yes ☐ *No* ☐
nurses? N/r ☐ Yes ☐ *No* ☐
pharmacist? N/r ☐ Yes ☐ *No* ☐
programmer? N/r ☐ Yes ☐ *No* ☐
statistician? N/r ☐ Yes ☐ *No* ☐
monitor? N/r ☐ Yes ☐ *No* ☐
12.2 Are payments to collaborators
adequate? ☐
excessive? ☐
or *inadequate?* ☐
12.3 Is the equipment to be ordered
adequate? ☐
excessive? ☐
or *inadequate?* ☐

12.4 Are the costs of overheads, travel, and so on

adequate? ☐

excessive? ☐

or   *inadequate?* ☐

A13 Supporting documents

13.1 Are the trial documents clear? ☐

or *ambiguous?* ☐

13.2 Are the trial documents

efficient for data-processing? ☐

or *inefficient?* ☐

13.3 Is the information collected *too much*? ☐

sufficient? ☐

or *too little?* ☐

13.4 Are the methods of measurements well

described or not?   Yes ☐ *No* ☐

13.5 Are they good methods?   Yes ☐ *No* ☐

13.6 Will quality control be

nonexistent? ☐

mediocre? ☐

or efficient? ☐

A14 Summary

14.1 Does the summary do the trial justice?   Yes ☐ *No* ☐

14.2 Is the trial ethical?   Yes ☐ *No* ☐

14.3 Will (would) you fund this trial?   *No* ☐

Possibly ☐

Yes ☐

9.   RECRUITMENT OF SUBJECTS

The recruitment of subjects must be considered very carefully to ensure that sufficient persons with the characteristics required by the investigators are enrolled within an appropriate period of time. The number of subjects is of great importance and was discussed in chapter 6.

SUBJECTS WITH THE REQUIRED CHARACTERISTICS

General selection criteria

Selection criteria are discussed in chapter 3. These criteria are required to ensure that the subjects have the condition being investigated in the trial and that they cannot be expected to experience an adverse event as a consequence of entering the trial. The general selection criteria will usually include the sex, age, and race of the subjects.

Criteria not usually defined in the protocol

The type of patients recruited will greatly affect the extent to which the results may be generalised (chapter 13). The protocol may not state the social class of those to be recruited and there has been considerable anxiety in the United States that clinical trials are conducted on *"people who are least likely to complain or who are least likely to have the power to make their objections felt"* [195]. On the other hand, Jeremiah Stamler reported that in the Hypertension Detection and Follow-up Program (HDFP), *"we elected to have a sizable group of (black) patients from the slums of Baltimore, Md., Birmingham, Ala., and Washington D.C."* [196]. When recruiting patients it may be important to consider social class and other features relevant to generality, and the recruitment policy for the HDFP trial allowed the outcome to be compared for whites and blacks.

THE PERIOD OF RECRUITMENT

The period of recruitment is often not stated in advance and this may have severe disadvantages as the interval tends to become prolonged. Fast recruitment reduces the total length of the trial, the costs involved, and the difficulty of keeping the investigators' enthusiasm at a high level. Even when recruitment times are fixed in advance for a given number of patients, these times may easily be doubled. For example, recruitment to the LRC Coronary Prevention Trial increased from one to two and one-half years [197]. In the Aspirin Myocardial Infarction Study (AMIS) recruitment was limited to one year but was slow at first with a sharp rise prior to the deadline. *"Doctors proved to be crisis orientated and most of the recruitment came in the last quarter of the recruitment phase"* [198].

METHODS OF RECRUITMENT

Patients have been recruited for randomised controlled trials from the investigator's own medical practice and from medical colleagues. Suitable patients have also been identified from medical records, from screening programs, and even from volunteers responding to advertisements.

From the investigator's own practice

Patients are most frequently selected from the investigator's own practice. This method may be very effective when only small numbers of patients are required but is rarely applicable to large-scale trials.

The investigator will already be known to the patient and hold a position of trust. This relationship may render the patient anxious to take part in a trial in order to please the investigator and great care has to be taken not to coerce the slightly unwilling patient into taking part.

Referrals from other medical colleagues

The details of the trial must be explained to colleagues who are not taking part in the trial and despite a considerable amount of effort, very few referrals may be forthcoming [197,198]. Independent clinicians may not be motivated to refer their patients and may forget about the existence of the trial. When practitioners receive fees for items of service, referral of patients to the trial investigators may deprive the referring clinicians of financial income for the duration of the trial because many trials provide patients with free attention and medication.

Identification of possible patients from medical records

Physicians' notes and laboratory records have been examined in order to identify suitable patients.

Physicians' notes
Physicians' records were examined in the AMIS trial of secondary prevention of myocardial infarction. The notes usually gave the entry criteria of age, sex, and electrocardiographic and enzyme changes so that suitable patients could be selected. Schoenberger [198] reported that this was a better method of recruitment than physician referral and the medical profession was not opposed to this method of selection. However, only 10-20 percent of those patients invited to take part actually entered the trial.

Laboratory records
In the LRC Coronary Prevention Trial, patients with type II hyperlipidaemia were entered into the trial. Having failed to achieve many physician referrals, the investigators asked commercial laboratories for suitable patients [197]. The laboratories appeared to divide into three groups.

1. One laboratory wanted payment for the names (this laboratory was not used).
2. There were laboratories who felt a breach of confidence could occur but who agreed to write to the physicians caring for the identified patient to suggest referral for the trial.
3. One laboratory allowed the trial staff to review the results and write directly to the physician.

In the LRC trial both physician and laboratory referrals led to the entry of very few patients.

Screening to detect suitable subjects

Having recruited only 11 patients in Baltimore the LRC Coronary Prevention Trial employed screening methods to search for male patients with elevated serum cholesterol concentrations. These methods are given in table 9-1. Screening of men attending public gatherings; living in Columbia; being considered for another trial; or donating blood for the Red Cross led to the recruitment of a further 246 subjects. The screening required a recruitment coordinator, a blood-drawing team, and contact with the news media and film makers for television in order to advertise the activity.

Volunteers with the disease under study (self-referral)

Many patients entered into the AMIS trial referred themselves for inclusion in the trial. Patients were eligible for this trial if they had sustained a myocardial infarct in the preceding five years and volunteers were sought by the advertising methods in table 9-2 [198]. The patients' doctors supported the trial and over a third of those randomised were recruited in this fashion. However, a high proportion of volunteers were not suitable; some of these were excluded when they were first interviewed by telephone.

Table 9-1. Methods of screening men for the LRC Coronary Prevention Trial in Baltimore [197].

A. *Screening at:*

  1. Health fairs
  2. Before-church groups
  3. YMCA meetings
  4. Shopping centres
  5. Baseball games
  6. 57 Baltimore industries
  7. American Heart Association local programs

B. *Screening of:*

  1. The entire city of Columbia, Maryland
  2. Blood from American Red Cross donors
  3. Potential candidates from the Multiple Risk Factor
     Intervention Trial (MRFIT)
  4. Volunteers responding to mass newsletter mailings

Table 9-2. Methods of getting self-referrals for the AMIS trial [198]

---

1. *National Media Coverage*
2. *Locally*
     Mass mailing in utility bills
     Public rallies
     Radio and TV announcements
     Newspaper articles
     Paid advertisements

FACTORS RELATED TO RECRUITMENT

Recruitment may be increased by reducing the exclusion criteria, offering more financial support to the investigators, or threatening to withdraw funding. The principal investigators also have a great responsibility in generating enthusiasm by visiting, lecturing, and publishing preliminary information.

Changes in protocol

Making the entry criteria less stringent can increase recruitment. For example in the National Co-operative Gallstone Study recruitment was increased by raising the upper age limit from 69 to 79 [199].

Financial

In the Gallstone Study recruitment increased after it was threatened that research contracts would not be renewed. Similarly, in the AMIS trial contract support was reduced for centres who did not recruit as many patients as they should, and centres who recruited an excess were given increased contract funding.

Threats to withdraw a centre from the trial

In the AMIS trial all centres had to achieve a critical number of patients to remain in the study. It is not known whether increased recruitment in clinics who were stimulated to achieve the critical number compensated for the number of patients lost in clinics who were removed from the study.

FACTORS NOT RELATED TO RECRUITMENT

Croke has reported some factors that did not influence recruitment to the various centres in the National Co-operative Gallstone Study [199]: population density, frequency of performing cholecystectomies, expertise of the clinical directors, and incentives provided by ancillary studies were not related to recruitment rates.

CONCLUSIONS

Recruitment is always a major difficulty in clinical trials. Muench's third law states, *"the number of patients promised for a clinical trial must be divided by a factor of at least 10"* [200].

In large trials recruitment may occur from clinical practice, laboratory records, population screening, and volunteers with the appropriate disease. Recruitment may be increased by widening the entry criteria and by financial incentives.

Richard Peto, speaking at a meeting in Lyons in November 1981, emphasized the importance of recruiting sufficient patients for trials in cancer patients. He suggested that 1,000 patients are usually the minimum number required and that recruitment can be increased by collaboration between centres, simplification of entry and follow-up procedures, and possibly, by payments to compensate for secretarial expense. The ISIS and GISSI trials are a consequence of these methods in the cardiovascular (post infarction) field. The first three ISIS trials recruited over 16, 17 and 41 thousand patients by employing very simple entry and follow-up procedures [77,201-203].

10. INFORMATION TO BE COLLECTED DURING A TRIAL

Information must be recorded before a subject enters the trial, throughout the course of the trial, and at the end. Before the start of a clinical trial it must be documented that the patients have the condition under investigation, that there are no contraindications to their entering the trial, and that informed consent has been obtained. During the course of the trial both the benefits and adverse effects of treatment must be recorded to demonstrate that the patients may safely continue in the trial. At the end of the trial the final data must be recorded. These will include full details on defaulters and the attempts made to contact them. This chapter also considers the quantity of data to be collected, the design of documents, the questions to be asked, and the various stages of data preparation.

THE QUANTITY OF DATA TO BE COLLECTED

Hamilton has cautioned against collecting too much information: *"It would be better to resist the temptation to collect every kind of information and spend the time first in thinking more carefully about what would be relevant, and to devise hypotheses to be tested"* [204]. Wright and Haybittle have agreed with this assessment and cautioned that the investigator may be unwilling to enter a patient in the trial if there is a lot of paperwork and also that the quality of the paperwork may deteriorate as the quantity increases [205]. On the other hand, data have to be carefully recorded on the outcome of the trial, both on the benefits and disadvantages of treatment; the initial information must be adequate to identify the patients in the trial and to show that the different treatment groups were similar with respect to important starting characteristics. If minor objectives have been defined these will also require the collection of extra information.

FACILITATING DATA ENTRY

Three strategies will reduce the time spent on data recording by the individual investigator. The patients or ancillary staff may complete some of the documents, and some information may possibly be transferred from one computer to another.

The patient can complete certain documents

The patient can be asked to complete a form giving all the identifying and demographic information discussed below. This form can be extended to include past medical history, past treatment, family history, occupation, cigarette and alcohol consumption, and other items relevant to the trial.

During the course of the trial the patient may be asked to complete self-administered questionnaires on symptomatic and general well-being (chapter 15). This strategy may save the investigator time and effort and may improve the quality of the data when subjective symptoms have to be assessed.

Ancillary staff may complete certain documents

Clerical, secretarial, or nursing staff may prove more accurate and conscientious than the investigator when transcribing investigation and other results from the medical records to the trial documents. Whenever possible these tasks should be completed without interruption by the urgency of patient consultations.

Direct transfer of computer-held information to the computer-held records for the trial

Biochemical results, electrocardiographic tracings, and haematological findings are often held on computer and, in theory, could be transferred directly to the trial database for final analysis. However, differences in computer software used may lead to formidable data conversion problems. The details of patient identification may also vary between the different sources and pose extra problems when extracting information from one computer file and including it in another. However, in view of the real risk of transcription and data entry errors, this form of data collection is highly desirable, where possible.

INFORMATION IDENTIFYING THE PATIENT

The record forms used in a trial must be considered as confidential information but in a clinical trial the degree of confidentiality need not be greater than with the usual medical records. It is therefore common practice to include name, address, and other identifying features on the first document to be completed. Subsequent record forms may include extremely confidential information or be sent through the post with the possibility that they are opened by the wrong person or discarded in a public place. For example, a symptom questionnaire has been mailed that included questions on sexual function in order to detect drug side effects [206,207]. It is prudent to identify these documents with a trial number alone, and the code should be held by the investigator. The patient should be assured of the confidentiality of the information; this is important as documents have gone astray in hospital postal systems and been discovered in discarded refuse (hopefully after the data have been abstracted). The investigator must take great care of all confidential information and the record forms stored carefully and be shredded before final disposal.

If the name and address are only included on the initial trial record form and a code number used thereafter, the use of a single number alone may lead to errors in that the wrong number may be entered on a particular record form. For this reason and when sensitive information is not being recorded, many investigators prefer to have both name and trial number on every document. Similarly, two separate numbers for identification will also limit any difficulty in identification.

Items required for identification

Items required include: (1) full name; (2) address; (3) trial number; and (4) hospital number when appropriate.

Other demographic information required

Other information required includes: (1) sex; (2) date of birth; and (3) race.

Demographic information that may be required if the patient is to be identified in central government records

If a patient in a long-term trial is lost to follow-up it is essential to determine whether or not he has died. In the United Kingdom the Office of Population Censuses and Surveys can inform a bona fide medical research worker, for a small fee, whether a patient is dead or alive and if the person is dead, the office can provide a copy of the death certificate. The tracing of such a patient will be facilitated if the National Health Number of the patient has been documented. Also useful for this purpose is the marital status of the patient, maiden name, the name of the patient's primary care physician, place of birth and last address.

Other necessary information

Telephone number of patient
The patient's telephone number must be obtained because the patient may have to be contacted quickly.

Marital status of the patient
This will enable female patients to be addressed correctly but will only rarely be relevant to the conduct or results of the trial.

Name and address of the patient's primary care physician
When the trial is not being conducted by the primary care physician he must be informed of the details of the trial. In the United Kingdom every patient has a general practitioner and his agreement should be sought for the patient to take part in a trial. He must be given the opportunity of objecting to his patient's taking part although his written consent is not usually necessary. In my experience the cooperation of the primary care physician can be of great assistance in a long-term clinical trial. Glaser considered that this cooperation is essential in trials involving employees in a pharmaceutical company as the family doctor may be aware of a condition or circumstance that makes it inadvisable for the subject to take part in an experiment [32].

DATA TO SHOW THAT THE PATIENT IS ELIGIBLE FOR THE TRIAL

Prior to a patient's entry into the trial it is important to document that he satisfies all the entry criteria and has no contraindication to participation in the trial. These criteria are discussed in detail in chapter 3. It is essential that the trial records confirm that all exclusion criteria are in fact negative and all inclusion criteria positive. Quality control cannot be carried out effectively without this documentation.

DOCUMENTATION OF INFORMED CONSENT

Normally the signed consent form will be included with the trial records; often a copy is given to the patient. If written consent was not required then a note must be made in the trial documents that verbal consent was obtained on a certain date. If a third party was present at these discussions, this fact should be noted.

DATA TO BE RECORDED DURING THE TRIAL

As discussed above, it is very difficult to decide how much information should be recorded during the course of the trial. Data relevant to the prime objective of the trial have to be fully documented. Subsidiary objectives are usually identified and data relevant to these must also be recorded. It is also essential to record adverse drug effects and symptom side effects.

DATA TO BE RECORDED AT THE COMPLETION OF THE TRIAL

At the end of the trial the most important end points are determined: for example, the fasting serum cholesterol in a trial of lipid-lowering treatment. If a patient does not appear for this final visit or any other trial consultation, he must be contacted quickly and arrangements made for him to be seen as soon as possible. In a trial of drug treatment it may be possible to give the patients an extra supply of tablets so that if they miss one visit they can continue with treatment until seen. It is of the greatest importance to trace defaulters as they may have died or been withdrawn owing to some adverse effect of treatment or, if in a control group, due to an adverse effect of inadequate treatment. Such patients must be contacted and asked to return to see the investigator in order for the reasons for default to be determined. The trial documents must include all the details on these patients including the methods used to recall them, the reasons for default and in the event of death, the date and causes of death.

THE DESIGN OF THE DOCUMENTS

The investigator must be able to complete any forms quickly and easily. The questions must be clear and unambiguous so that the answers recorded are correct, and the documents should be designed so that the data can be processed accurately and efficiently. The investigator should consider the layout of the forms, the provision of instructions to those who complete them, and the duplication of the materials for reasons of security.

Layout of the forms

All documents should be headed by the name of the institute in which the trial is being performed. If there is any possibility of the form going astray, the name and address of the investigator should also appear. When the form is to be completed by the patient it is important that a brief note accompanies the questionnaire or be printed on it stating who is asking the questions, why they are being asked, and assuring that the answers will be treated with complete confidentiality. The patient must also be instructed how to complete the form (see figure 10-2, p195).

The first items on the documents should be those collected first, as the forms should be completed in the order in which the data are obtained. For example, in a trial of a hypoglycaemic drug, details of previous treatment will be available first and entered first, results of clinical examination will be available next, and the results of investigations, such as blood tests, will be available later and entered on the end of the form.

When one investigator is completing his own trial documents, presumably he will not be concerned with strategies that increase the proportion of questionnaires that are completed. When patients or many different investigators have to complete the documents, it is important to improve the layout in order to maximise response. The reader is referred to standard texts on this subject [208,209] and a very relevant book entitled 'Data Collection Forms in Clinical Trials' by Spilker and Schoenfelder [210]. This book provides sample forms suitable for the entry of neurological and other information, clinical chemistry and haematology and patient termination forms.

The investigator may not be able to compete with the postal sales techniques to increase

In the last 3 months have you suffered from <u>attacks</u> of the following:

sweating	☐ No	☐ Yes
palpitations	☐ No	☐ Yes
confusion	☐ No	☐ Yes
severe apprehension	☐ No	☐ Yes
hunger	☐ No	☐ Yes
sleepiness	☐ No	☐ Yes

Figure 10-1. Section of a questionnaire given to diabetic patients [211]. The condensed format led to some patients indicating only positive information and omitting negative answers.

response, for example, where the recipient is told that he has won a prize. To receive this gift he has only to stick a 'yes' stamp into an exactly matched space and return the document in a postage-paid envelope. (In order to avoid purchasing an additional item, he may have to take a psychologically less attractive action, for example, refusing an additional and generous, but not free, offer from his benefactor.) Sales techniques have not yet been widely employed in randomised controlled trials, but if the investigator wishes a questionnaire to be completed, he should consider the aesthetic layout of the document, provide postage-paid envelopes, and consider whether the questions should be answered by ticking yes or no boxes rather than deleting the incorrect answer or entering a correct answer in freehand.

To avoid errors, a box to be ticked should be sited very close to the appropriate answer.

A common source of error is to shorten the questionnaire by amalgamating questions into blocks as in figure 10-1 and not insisting on an answer to each individual question. In theory the phrase "In the last 3 months have you suffered from *attacks* of the following" does not have to be repeated for every question but patients may not answer all the questions in such a sequence, often only indicating the positive responses. A longer section asking each question separately would result in a higher completion rate for the individual questions and avoid having to make the assumption that no response equals not present.

Instructions to person completing the form

The instructions for completing a form should appear on the document to be completed and not on a separate sheet as the latter may not be consulted. When appropriate, the instructions should be adjacent to the question being asked and should state how the responses are to be recorded. For example, if the correct answer has to be ticked, a cross may cause confusion [205].

Duplication of forms prior to use

A small trial may only involve word processing the forms and photocopying the top copies. However, in a larger trial many documents may be required and printing will have to be considered as this has advantages over photocopying. Wright and Haybittle [205] considered that printed words are easier to read than typescript as typewriters do not allocate characters a space commensurate with their size but give each letter equal space. In this respect care has also to be taken with the printing from word processors. These authors also advised a print size comparable to newspaper text and they warned against the use of capitals alone rather than both lowercase and capital letters. The use of capitals on their own can increase reading time by 12 percent [212]. A larger font size is required if the documents are to be completed by elderly subjects.

Photocopying or printing will provide the forms to be completed but there remains the problem of duplicating the documents after use. No-carbon-required paper may be useful.

The use of no-carbon-required paper

The documents in a clinical trial are of great importance and the use of no-carbon-required paper will allow records to be made in several copies. The copies should be filed separately as an insurance against damage by fire or water, loss in the post, or theft. The loss of research data by theft will probably be accidental, but imagine that your only set of trial documents might be stolen along with your car! Great care must be taken of the information. I well remember the fate of some data I had laboriously collected and recorded with a fountain pen containing washable ink. Someone left a tap running over a weekend in a laboratory on the floor above my office and, following the deluge of water through the ceiling, the data were more than difficult to interpret! Collect all data in duplicate or triplicate and store the copies separately.

THE FORM OF THE QUESTIONS AND ANSWERS

The characteristics of a good question for a patient are discussed in chapter 15. The present section considers the most desirable features for questions in a trial document to be completed by an investigator: namely, lack of ambiguity; use of positive terms; ease of comprehension; and necessity for a single response. There is obviously a considerable degree of overlap between this section and chapter 15.

The question and answer options must not be ambiguous

Wright and Haybittle provided a good example of question and answer options [205]. In response to a question on the size of a tumour mass, an answer option was: *"Reduction of tumour diameter by less than 50%."* The authors suggest the answer would be better phrased as, *"smaller but not as small as half the original diameter."* The original answer is a phrase that is not strictly ambiguous but simply difficult to understand.

The question should use positive terms

Clark has shown that positive terms are easier to understand than negative terms [213]. For example, the question, *"Is the right first toe longer than the right second toe?"* is to be preferred to the question, *"Is the right first toe shorter than the right second toe?"* Short persons may not agree!

The question must be easily understood

The question must not include difficult words and must be understood by the least intelligent observer. Never use a long word when a short one will do. Abbreviations should be avoided if they are not familiar to all investigators and the questionnaire should also be grammatically correct.

The question must only require a single response

If an observer is asked to record poor circulation to an extremity, he can be asked, *"Is a hand or foot white, blue or cold?"* The answer option is yes or no. However, the investigators may regret that they did not record whether it was one hand, both hands, a hand and a foot, or even whether only certain fingers or toes were affected. Similarly they may wish to know whether the affected extremities were white, blue or cold. A series of questions leading to single responses would be preferable. Alternatively, the investigator can be asked to list the fingers or toes that are white and, similarly, those that are blue and those that are cold. If the question is crucial to the outcome of the trial, the temperature of each extremity should be recorded, as an objective measurement is usually to be preferred to a subjective assessment, assuming the objective measurement is both valid and repeatable.

THE RESPONSE TO THE QUESTION

The observer should be asked to tick a response rather than enter a coded reply

An investigator may be asked to indicate whether the patient is taking an anxiolytic drug or not and, if so, which one. He could be asked to enter the name of the drug, but alternatively he could be asked to tick the name on a complete list of possible drugs or to examine the International Nonproprietary Names for Pharmaceutical Substances Classification [214] and enter the numerical code provided by this classification. The provision of a complete list of drugs will enlarge the document considerably but will have the advantage of reminding the investigator of the names of antianxiety drugs.

There are two problems in entering a numerical code from the international classification. First, the form may have to be completed in a hurry in a busy clinic with no time available for the document to be consulted; second, a physician-recorder, who rarely performs coding duties, may be less accurate than trained clerical personnel whose main job is to provide this service.

In general, the investigator should be asked to enter the response in freehand for subsequent coding or to tick a list of options rather than enter a coded response.

Enter responses in preprinted boxes

The provision of boxes may or may not improve the completion rate, but they should be used for numerical data in order to facilitate later entry to computer files. It has been shown that entering data directly into boxes may increase writing time by ten percent and reduce legibility by three percent [205]. However, when the data are subsequently converted to a machine-readable form time will be saved and some errors may be avoided. The boxes also indicate where an answer is expected.

The boxes must be large enough for those with big handwriting; where appropriate the units should be printed and the boxes segmented to the correct number of figures (figure 10-2); and for biochemical data, the position of the decimal point indicated. The instructions should also be clear and indicate how to mark the boxes, how to enter the figures (figure 10-2), how to make corrections or record missing information and for the date, the order that the day, month and year have to be entered. Instructions on making corrections are important if the data are to be read using an optical scanner and instructions on missing information are important to differentiate a missing value from a 'zero' or 'not present' response.

Please tick (√) correct box or enter required number. All information will be treated with complete confidentiality.

Trial Number | 0 | 1 | 4 |

Is your father still living? No [ ] Yes [ ]

If 'Yes' how old is he now? [ ][ ][ ] Years

(Please enter one figure in each box,

eg | 0 | 5 | 0 | for fifty years)

If 'No' how old was he when he died?
[ ][ ][ ] Years

Figure 10-2. An example of part of a questionnaire, designed so that subsequent computing is facilitated.

Order of yes and no answers

When several yes/no answers are required consecutively it is advisable not to have a particular response always stated first as it has been suggested that some respondents prefer to tick the first box irrespective of its contents. Occasionally reversing the position of yes/no answer options may inhibit a repetitive response but may lead to errors when most first answers are, say, yes and suddenly one is no. The no may be ticked in mistake for a yes. The best solution may be to roughly alternate the yes/no and no/yes answer options.

CODING THE RESPONSES

The information provided on the trial documents will be analysed by computer to limit the errors in calculation and allow complex statistical computations to be made. If the arithmetic tasks to be performed are not complex, analysis by computer program may take longer than the use of a calculator as the data have to be read into the machine. However, the investigator usually attempts so many analyses - for example, of many variables in several subgroups - that the use of computer programs is inevitable. The data must be transferred to computer via a visual display unit and an attached keyboard or scanned by an optical scanner. The data must be coded to a numerical format and identified as requiring transfer to a computer file. The position that the data must occupy in that file must be known.

The information may be keyed directly from the questionnaire

Computer hardware and software change very quickly. At the present time programs exist so that an image of the trial document appears on the screen and the data entry personnel enter all information enclosed in boxes or otherwise marked for entry. The modern systems have done away with the necessity for a separate hand coding of information onto special computer entry forms. The operator must be aware of what to enter through the keyboard. For example, if the gender box 'M' is indicated, an 'M' should usually be entered but older systems may require '1' to be entered.

Most data entered through this system have to be double-entered to detect key punching errors. Two different operators enter the same data on different occasions and computer programs identify differences. Errors such as entering 'F' instead of 'M' should be very low and probably less than 1 in 1000. Poor writing, where a 'O' resembles a '6' will lead to a higher error rate.

The information may be entered by optical scanning

In this instance the scanner identifies a mark and enters the information. Therefore a cross in the 'M' box will lead to a record of being male. Optical scanning is necessary for entering large amounts of qualitative data. Problems may arise if the forms are not accurately printed and if corrections have been made to the questionnaires. It is therefore necessary to visually examine the forms prior to entry to detect comments and alterations which the scanner will not interpret. Optical scanning is less useful for a long set of variable numeric data, such as biochemical results. These still have to be keyed in and double entered.

The data must be checked

Computer programs exist to detect out-of-range information and these can trigger a message at the time of entry. Moreover consistency checks can be set up to detect impossible situations such as a pregnant male or a systolic blood pressure lower than a diastolic pressure.

CONCLUSIONS

This chapter has outlined the information that should be documented for an individual subject, before, during, and after the performance of the trial. A compromise has to be made between collecting all the information that could conceivably be of interest and limiting the effort involved by documenting only the most essential data. The design of the documents was

discussed, including their layout and the form of
any questions. The methods of computer entry for
trial data were introduced, and strategies to
facilitate data entry and subsequent processing
were discussed.

11. THE CONDUCT OF THE TRIAL - GOOD CLINICAL PRACTICE

This chapter considers some of the practical difficulties in performing a trial and the problems that may arise. These may be divided into those that are foreseen and taken into account of prior to the start of the trial and those that are unexpected. Murphy's law states, *"If anything can go wrong it will."* This law has been attributed to captain E Murphy, a development engineer, who applied it first to an individual technician saying, *"If there is any way to do it wrong, he will"* [215].

Owing to the probability of unforeseen difficulties it is desirable to have either a separate pilot trial or to commence the trial in a pilot fashion using a provisional protocol. If the pilot trial proves satisfactory and the protocol only requires minor modifications, the trial may continue; otherwise the initial design is abandoned as impracticable. After considering the pilot trial, this chapter discusses the use of a run-in period, the problem of noncompliance by the subject or the investigator, Good Clinical Practice (GCP) and the difficulties that arise in stopping a trial.

THE PILOT TRIAL

It is important that the protocol and design for a large or long-term trial is tested either in a separate pilot trial or during a preliminary period of pilot running during which the protocol is open to amendment. The performance of a pilot trial allows the technique of measurement and treatment to be tested, the optimal treatment schedule to be determined, the administration of the trial to be tested, and the rate of recruitment to be assessed. Also, a preliminary estimate of the treatment effect may be obtained and the wider implications of the trial may become more clear.

The methods of examination and therapeutic techniques are tested

During the pilot trial an examination procedure may prove to be unsatisfactory and can be modified. Similarly a therapeutic regimen can be altered if necessary. In chapter 13 the alterations are discussed that proved necessary in order to measure visual acuity during a trial.

The optimum treatment schedule may be identified

The pilot trial may lead to a modification of the dose schedule for a drug by revealing that a large dose produces an unacceptable excess of side effects. Alternatively the pilot trial may demonstrate that an insufficient dose of the drug is being administered. This can be rectified subsequently in the main trial. When two active drugs are being compared in a long-term trial the pilot trial may confirm that they are being used in equipotent doses with respect to their short-term effects. For example, if two drugs are being compared for the prevention of gout over two years, it will be sensible to select doses that reduce the serum uric acid by a similar amount.

The administration of the trial can be tested

The early stages of a trial will involve the entry of patients, randomisation, and the provision of treatment. A coordinating centre may be responsible for the processing of trial documents, randomisation, and the distribution of drugs. All these administrative functions may be tested in the pilot study.

The rate of recruitment can be assessed

Muench's third law states, *"In order to be realistic, the number of cases promised in any clinical study must be divided by a factor of at least ten."* The law has two important corollaries: *"the length of time estimated as necessary to complete a study must be multiplied by a factor of at least ten"* and *"the sum of money estimated as necessary to complete a study must be multiplied by a factor of at least ten (without inflation)"* [171]. There is more than a grain of truth in Muench's third law and the pilot trial may provide an assessment of the recruitment rate that can be achieved in the main trial.

A preliminary estimate of the efficacy of treatment can be made

Although a trial will be designed to detect a certain magnitude of effect, presumably based on prior information, the pilot trial may be useful in confirming that this effect can be achieved under the conditions of the trial. A pilot trial cannot be expected to detect a given reduction in mortality or morbidity as a long period of observation may be required to detect such an outcome. However, if the trial is intended to alter a risk factor by a given amount with a view to subsequently altering mortality, the change in the factor can be confirmed in the pilot trial.

The pilot trial may enable the investigator to consider the wider implication of the study

Glaser has stated. *"A lesser study of an important question is usually of more value than an excellent study of a trivial question"* [32]. The reader may not agree with this opinion, but in a pilot trial the investigator may realise that more important questions could be answered by the trial. He or she may adjust the protocol accordingly to include more subjects or observations than initially envisaged.

The disadvantages of a pilot trial

A pilot trial may delay the start of the full trial, but this may be overcome by starting the major trial with a period of pilot running. This preliminary investigation may not lead to any major change in the trial protocol, but a minor modification may be most important and occasionally the initial trial protocol will be abandoned. However, the results of a pilot trial may be misleading. Muench's first law states, *"No full-scale study confirms the lead provided by a pilot study."* Other criticisms of pilot trials have been made including, *"Studies can be called pilot studies and so avoid criticism for poor design"* and *"pilot studies are a waste of time, money and effort"* [171]. It must be emphasized that the results of a pilot study must be treated with great caution; after all, they are only an initial reconnaissance expedition. However, pilot studies do lead to important changes in protocol, are often invaluable, and are always to be recommended. If the protocol is not altered in any major respect the results obtained in the pilot trial may be incorporated in the final analysis, provided an adjustment for repeated looks is made when necessary (chapter 6).

THE RUN-IN PERIOD

The run-in period for a trial is a period of observation during which subjects are considered for entry to a trial of treatment of a chronic condition. At the end of the period they are randomised to a treatment group if they prove to be eligible. A run-in period is not obligatory and certain advantages and disadvantages are given in table 11-1, and discussed in the sections below.

A run-in period may be appropriate if a preliminary period of observation is required to prove that the patients have a certain condition or to establish other baseline information. The time spent in the run-in period may also be used to: allow any effect of a placebo to take place and thereby reduce this effect after entry to the trial; detect noncompliance with therapeutic advice and exclude such patients; and allow the patients time to consider whether or not they wish to take part in the trial.

Table 11-1.   The advantages and disadvantages of having a run-in period prior to randomisation in a trial of treatment in a long term condition.

Advantages	Disadvantages
1. Diagnosis can be ascertained with greater certainty.	1. Patients in the trial are less typical of the general population of patients.
2. Allows baseline measurements to be made.	2. Dropout during the run-in period will increase the total number of defaulters.
3. Placebo response may be reduced after randomisation.	3. Ethical problems.
4. Some noncompliant patients may be identified and excluded.	4. Expense.
5. Dropout rates are reduced after randomisation.	

Advantages of run-in period

1. Establishing that the patients have the condition under investigation

If a patient is only examined on a single occasion, it may be difficult to establish a particular diagnosis with confidence. For example, it may be necessary to repeat measurements on subsequent occasions to prove that the patient has, say, a consistently high blood pressure, fasting serum cholesterol, or blood sugar. In addition, it may be necessary to order further investigations to prove that the condition in not secondary to a pathological process that requires immediate attention and that would exclude the patient from the trial.

2. Establishing baseline measurements

Before commencing treatment in the trial, baseline measurements may have to be made and this period of investigation can constitute a

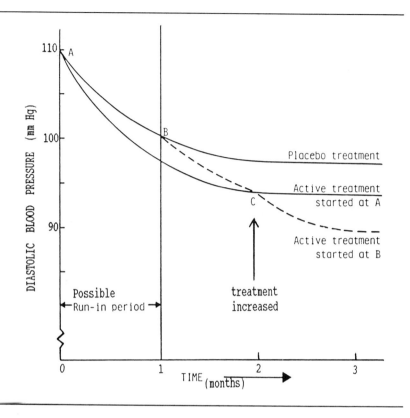

Figure 11-1. A hypothetical trial of the effect of an active treatment and placebo on diastolic blood pressure. When active treatment was started at A, a 15-mmHg fall in pressure was judged satisfactory at C and the dose of active treatment was not increased. When active treatment was started at B, following a run-in period, a 5-mmHg fall was not judged satisfactory at C and active treatment was increased (---).

run-in period for the trial. The establishing of a stable baseline is particularly important when regression to the mean is a problem (chapter 13). Failure to establish a baseline measurement is especially important when a placebo effect is present.

3.   **Estimating the placebo effect**

During the run-in period it may be appropriate to give a placebo in order to assess the response to this treatment. If there is a large response to placebo, this may mask the effect of active treatment. For example, in some trials the dose of a particular treatment is not fixed and has to be increased in a stepwise manner according to the response. This titration or increase in treatment may be difficult if a large early placebo effect is occurring. This situation is illustrated in figure 11-1 where active antihypertensive treatment has to be compared with placebo and measurements of blood pressure are presented in a schematic fashion for a patient whose initial casual diastolic pressure is 110 mmHg, falling to 100 mmHg after one month on placebo. This fall is due to a combination of regression to the mean, becoming accustomed to the measurement, and a possible effect of a placebo tablet. Figure 11-1 also provides the pressure when active treatment is started at A with no run-in period, and at B, after placebo run-in period of one month. When the drug is started at A, the investigator may be content with a 15 mmHg fall in pressure after two months of treatment. On the other hand, when active treatment is started at B the investigator is less likely to be content with a 5 mmHg fall after one month and may increase the treatment at point C. In this example, the presence of a placebo run-in period and increase in dose results in a wider separation between the effect of placebo and active treatment than when a run-in period is not employed.

Trials with a placebo run-in period and a titration of antihypertensive medication tend to reveal a larger fall in blood pressure on active treatment than trials without a run-in period. The EWPHE trial with a run-in period reported a difference between the active and placebo groups of 25 mmHg systolic and 12 mmHg diastolic [216], whereas the MRC trial without a placebo run-in period observed a difference of 15 mmHg systolic and 7 mmHg diastolic [181]. There are other differences between the two trials that may account for these results, but the MRC Working Party reported, *"Differences in mean pressure between treated and control ... were less than expected because the fall in pressure in the controls was greater and more prolonged than expected."* It is important to start the trial after any nonspecific treatment effect has terminated and the baseline measurement is constant.

4. Noncompliance can be determined

Noncompliance with therapeutic advice may be determined when a placebo is given during the run-in period (see below). Similarly the willingness of the patients to return for follow-up appointments is tested as well as their compliance with biochemical, radiological, and other investigations. Noncompliant patients may be excluded from entering the trial, but the trial population will be less representative of the whole population as a result.

5. Dropout may be reduced following randomisation

Default from follow-up during the main course of the trial is reduced following a placebo run-in period as some patients who are unwilling to return for repeated visits or investigations will be excluded. Also some patients become ill soon after starting placebo and attribute the illness to the placebo medication. These patients will usually be excluded or will exclude themselves from the trial. Many of these patients do not experience a coincidental physical illness but are worried about the trial treatment and experience a psychological reaction. Patients who react adversely to placebo medication (and therefore can be expected to react adversely to other treatments) may therefore be excluded. In addition, the run-in period gives the patients time to reconsider whether they are really willing to participate in the trial. Default appears to be more frequent in the initial stages of a trial and a run-in period excludes a proportion of those who would default after randomisation.

Disadvantages

A run-in period will reduce the degree to which the patients entering the trial are representative of the population as a whole. It increases the duration of the trial and may raise the overall default rate, increase the expense of the trial, and pose additional ethical problems.

1. The trial population will no longer be representative of all patients

Randomisation of subjects eligible for a trial will usually ensure that the different treatment groups in a single trial are similar in important characteristics. However, different trials will include dissimilar patients. Hampton has suggested that similar trials of secondary prevention in myocardial infarction may arrive at different results owing to the unequal characteristics of patients entering the different trials [217]. He pointed out that the mortality in the placebo group may provide a clue as to how representative the trial patients are of the population of patients as a whole since the usual survival of patients with myocardial infarction is known from prospective population-based epidemiological studies.

It is possible that the results of a trial will depend on the severity of the disease process, and it may be difficult to recruit patients who represent the whole community and have an average severity of the condition. Indeed, this may not be desirable if only patients with a given degree of disease severity are considered suitable for the treatment. A trial of a beta-adrenoceptor blocking drug in the secondary prevention of myocardial infarction [218] provided documentation both on patients who entered the trial and all patients considered for the trial. This information may be very valuable. Figure 11-2 gives the details for this multicentre trial. An unknown number of Norwegian subjects aged 20-75 years sustained a myocardial infarction, but about 11,000 were admitted to coronary care units during the period of recruitment for the trial. Of these, less than half met the stringent trial criteria for myocardial infarction and 508 died too quickly to

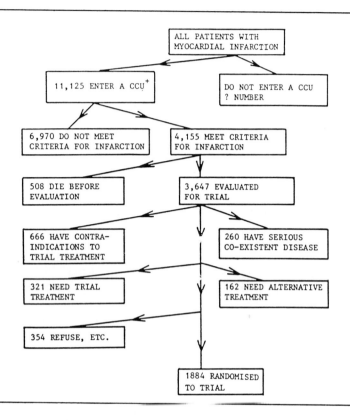

Figure 11-2. Recruitment to the Norwegian Multicentre Trial of Secondary Prevention in myocardial infarction using timolol [218].

CCU = coronary care unit. Some patients were admitted and evaluated more than once.

enter the trial. The remaining 3,647 were evaluated for the trial but only 1,884 were randomised. The reasons for not entering the trial were: contraindications to trial treatment; serious coexistent disease; trial treatment necessary; alternative treatment required; and administrative reasons such as the refusal of the patient. At the most optimistic estimate, less than ten percent of the population sustaining a myocardial infarct entered the trial. This shows how difficult it would be to recruit a representative sample.

A run-in period may make the trial subjects even less representative, as the information obtained during this interval may result in the exclusion of patients who do not have all the features of a disease or who are unwilling to attend regularly for supervision and investigation or who do not comply with therapeutic advice. The trial population, unrepresentative of the total population in the first place, is further narrowed to a group who are willing to take advice and medication and to put themselves at inconvenience for the benefit of medical research. This is acceptable as clinical trials are usually intended to examine the effects of treatment in those who receive the intervention. However, it is not surprising that sometimes the results of clinical trials may not apply to the population as a whole.

2. A long run-in period may result in an increased total number of defaulters

The run-in period may so increase the length of the trial that the default rate from the trial as a whole is increased, even when the dropout rate is reduced for the period of the main trial. However, dropout after randomisation is much more important than default during the run-in period. The latter only reduces the numbers available for the trial but default after randomisation may reduce the comparability of the different treatment groups. If the run-in period reduces the default rate after randomisation, this will be a major advantage (chapter 4). This advantage, however, will not hold if subjects are randomised prior to a run-in period.

3. Increased expense

A run-in period will prolong the length of the trial and add to the expense. If a placebo is to be taken during this period, then the cost of this medication and its administration must be taken into account.

4. Ethical problems with run-in period

It may not be possible to leave patients untreated or to give only a placebo for a period of time. This is discussed in chapter 3. When deciding whether or not to have a run-in period the investigator must examine the advantages and disadvantages for the trial under consideration. When one of the trial treatments is a placebo, and there are no ethical difficulties with this treatment, then a single-blind run-in period on placebo is usually desirable.

THE PROBLEM OF WITHDRAWAL FROM A TRIAL

Subjects may be withdrawn from a trial because they default or do not comply with the protocol. They may also die from causes unconnected with the trial or its treatment. The most important patients are those whose withdrawal is related to the treatment or trial end points.

At least four percent of patients taking part in a long-term trial of treatment of hypertension may default from follow-up every year despite attempts at recall [52]. The possible number of defaulters can be estimated before starting the trial and the numbers required for the trial can be adjusted accordingly. During the course of the trial it is very important to establish that the pattern of default does not differ between the treatment groups. When the reason for default or the proportion of defaulters does differ, omitting them from the analysis may bias the results; the problems of such an analysis are discussed in chapter 14. The results of such a trial must be analysed on an intention-to-treat basis (this is, without omitting defaulters or any other randomised subjects). The alternate analysis when subjects who do not follow the protocol are omitted has been called a per-protocol or on-treatment analysis.

The numbers required for a trial will increase accordingly to the proportion of defaulters. With a trial to be analysed on the per-protocol basis the increase will ensure sufficient numbers to demonstrate a given effect and with an intention-to-treat analysis the proportion of defaulters will indicate the extent to which the effect of treatment may be diluted.

Types of withdrawal

1. Withdrawals due to reasons unrelated to the end point of the trial

Subjects may be withdrawn when they move address, emigrate, or develop some condition that prevents their further participation in the trial (for example, a fractured spine following a road traffic accident). When calculating the numbers for a trial the number of dropouts must include an allowance for these withdrawals. Such withdrawals should not bias the results of the trial and should be equal in all groups. Any difference between an intention-to-treat and a per-protocol analysis should not be explained by these withdrawals.

2. Deaths from causes unrelated to the trial end point

The probable number of unrelated deaths can be determined from national statistics. In a long-term trial these numbers should be incorporated in the withdrawal rates, especially in trials of treatment in the elderly.

3. Withdrawal due to criteria related to the trial end points

Withdrawal for reasons related to a trial end point may pose great problems in analysis. The problem can be illustrated by trials of the long-term active treatment of hypertension versus placebo, with the end point of stroke events. We can consider two examples of withdrawals related to the end point: the development of an adverse drug effect to active treatment and a deterioration in the disease process while on placebo.

The development of an adverse effect of active treatment
A differential rate of adverse drug reactions between two drug treatments will lead to more withdrawals from one of the groups. This is most obvious when one of the treatments is a placebo. Moreover, it is possible that patients with

extremely mild or very severe disease are more prone to get an adverse effect to active treatment and be withdrawn. In the EWPHE trial [60] two treatments were employed in the actively treated group. Initially a combination of diuretics was given and if blood pressure control was not satisfactory, methyldopa was added. Therefore patients with severe hypertension in the actively treated group may be withdrawn owing to adverse reactions to methyldopa. In the placebo group severe hypertensives are less likely to be withdrawn as a result of placebo methyldopa treatment. However, there are other reasons why severe hypertensives may be differentially withdrawn from the placebo group (see below).

It is not always necessary to withdraw the patients because of an adverse effect. For example, hypertensive patients on diuretic treatment may develop diabetes mellitus. The trial protocol can stipulate withdrawal, antidiabetic treatment, or substitution of the diuretic treatment by another active antihypertensive drug.

Withdrawal of patients due to a deterioration in the disease process
A trial of antihypertensive treatment in the prevention of stroke may enter untreated hypertensives with a diastolic pressure in the range 85-94 mmHg. However, there may be evidence that patients with diastolic pressures higher than 100 mmHg require treatment. For ethical reasons patients who develop diastolic pressures greater than 100 mmHg must be withdrawn although they have not reached the important trial end point. Moreover, these patients will come almost entirely from the placebo group. As a result of such withdrawals the placebo group may contain progressively fewer severely hypertensive patients as the trial continues. The patients who are withdrawn cannot be said to have reached the stroke end point and yet cannot remain in the trial as it is believed that they may sustain a stroke. Randomisation should have provided groups that were similar at entry to the trial but selective withdrawal will destroy this equality.

The problem is so great that one group of workers who failed to observe a stroke in either placebo or actively treated groups wrote [121]:

"If withdrawal because of rising blood pressure is regarded as a unsatisfactory outcome then there were many more of these cases in the control than the treated group, but we are doubtful of the validity of this interpretation. Another consequence of these withdrawals is to alter the comparability of the groups. However comparable the two groups may have been at the outset, within 18 months this comparability had disappeared. In the present series the main function of treatment seems to be to prevent the rise of blood pressure, but it cannot be concluded from this evidence that treatment effectively reduces the risks of death or morbid events due to cardiovascular disease."

Methods for reducing withdrawals

The withdrawal or dropout of subjects from a trial can be reduced by a run-in period to detect patients who react badly to placebo or miss appointments; by not making the withdrawal criteria too stringent; by excluding patients who have a high probability of withdrawal; and by reducing the length of the trial. Withdrawals may also be reduced by maintaining close contact with the subjects during the trial, limiting demands on their time or patience, providing travel and other expenses, and continuing efforts at recall when they first miss an appointment.

Employ a run-in period
As discussed in the section above, the run-in period may be employed to reduce the number of defaulters after randomisation. The dropout rate appears to diminish with time and often subjects agree to enter a trial but do not return for their second visit. These early defaulters will be excluded during the run-in period and if a placebo is given, those who react with many symptoms and cannot take this medication will be excluded.

The withdrawal criteria must not be too stringent
In a trial lasting for, say, five years, it would not be reasonable to withdraw all patients who

missed two assessments or failed to take their medication for a two-week period. Absence of follow-up for six months or no trial treatment for a total of three months would constitute more reasonable criteria for withdrawal. Similarly, it may be reasonable to allow additional treatment to be given during acute illness without withdrawal even if this interferes with the effect of the trial treatment in the short-term. For example, an elderly patient in a long-term trial of antihypertensive medication could be allowed diuretic treatment during an episode of acute bronchitis and not be withdrawn from the trial, even though this therapy will lower blood pressure in the short term.

Exclude patients who may have to withdraw
Patients are often excluded from long-term trials if they have a serious coexistent disease that may limit their survival or ability to take part in later phases of the trial. Similarly, subjects who already know that they are going to move to another district, emigrate, or change their occupation are likely to default as the trial progresses; these subjects can be excluded from the trial.

The trial must not be too long
The longer the trial the greater the possibility that the patient will drop out or be withdrawn. The parallel-groups trial design has an advantage in this respect over the necessarily longer cross-over trial.

Miscellaneous factors
The subjects must be free to withdraw from the trial at any stage and cannot be asked to commit themselves to taking part for the whole duration. It is likely that the continued enthusiasm and interest of the same investigator may motivate the patient to continue. Factors that reduce default are not well documented, but the patients should not be kept waiting for long periods to see the investigator. They should be given their treatment and investigation without charge and be helped, where necessary, with the cost of transport to the trial centre.

What is the answer to groups made unequal by withdrawals?

Despite the measures discussed above, withdrawals will occur and the remaining patients be unrepresentative of the original randomised groups. During the course of the trial the withdrawn patients can be treated in two different ways. First, at the end of the trial they can be paired with patients in the other treated group(s) who are also 'withdrawn' from the analysis; second, the withdrawn patients can be considered to have reached an important positive or negative trial end point. At the end of the trial, withdrawal rates must be carefully analysed.

Withdrawal of similar patients from the other group(s)
We can consider again the example of active treatment producing diabetes mellitus. Patients developing this condition may be identifiable at entry to the trial (for example, from a high fasting blood sugar). If this is the case, for every patient 'removed' in the analysis from the active treatment group, a patient could be removed from the placebo group who was matched for initial blood sugar and other characteristics. This approach is full of difficulties as the initial blood sugar may not predict subsequent events and, if it is a good predictor, it would be better to exclude all patients with a high fasting blood sugar from entering the trial. The strategy of withdrawing similar patients from the analysis is not recommended.

Make withdrawal criteria important end points for the trial
Withdrawal rates must be carefully analysed and constitute an important outcome for the trial. Sometimes a withdrawal (or a proportion of them) may be assumed to be a cardiovascular end point. For example, if the incidence of stroke is known from previous observational studies for a particular level of blood pressure, a proportion of those withdrawn can be assumed to have had a stroke. Thus for a given number of withdrawals a number of end points may be assumed. To my

knowledge, the assumption that some withdrawals reach trial end point has not been utilised in the analysis of a trial, presumably as any conclusions would rest on unconfirmed assumptions. However, the results of the trial can be examined by assuming that the proportion of withdrawals leading to an end point is zero, one hundred percent, or various intermediate values.

Withdrawal in a cross-over trial

In a within-patient cross-over trial a patient having to be withdrawn from one treatment can proceed immediately to the next but with loss of data in the first phase of the trial. In a placebo-controlled trial the patients having an unacceptable response to placebo would be transferred to active treatment and those in the actively treated groups with an adverse effect would transfer to the placebo group. Although this is acceptable for a cross-over trial, the procedure cannot be recommended for a parallel groups design. However, the authors of the Co-operative Randomised Controlled Trial [121] suggested partial treatment for patients withdrawn from the placebo group owing to severely high blood pressure: *"One possible solution might be to maintain the diastolic pressures of the patients in the control group between 110 and 120 mmHg, or whatever upper limit of diastolic pressure was considered acceptable and compare the deaths and morbid events in this group with those in whom the pressure is maintained below 100 mmHg."* This strategy will compare full treatment in one group with partial treatment in another, hardly a satisfactory solution.

Withdrawal rates must be reported

The withdrawal rates should be presented for each treatment group and in different categories. Five useful analyses are total withdrawal; withdrawals due to death or morbidity unrelated to the trial outcome measures; withdrawal due to adverse drug effects; withdrawals initiated by the patient; and withdrawals initiated by other health professionals.

Conclusion on withdrawals

Patients may be withdrawn from a trial for
ethical reasons, they may refuse to collaborate
further or other health care workers may give
advice that leads to default from follow-up. The
default rate during a large clinical trial will
not be zero even over a short duration. In long-
term trials the default rate will be at least
four percent per annum and this fact should be
taken into account when calculating the numbers
required for such a trial.

The treatment may precipitate default and
defaulters must not be omitted from the analyses.
If analyses are conducted according to the
initial randomisation on the intention-to-treat
basis, the groups will maintain their initial
comparability. However, the interpretation of the
results may remain very difficult if withdrawal
results in a change of treatment. Also the
investigator may be interested in the effect of
treatment in a subset of the subjects (for
example, compliant patients). It is obviously
unsatisfactory to assess the effect of reducing
dietary fat intake on, say, blood cholesterol
when a proportion of patients refuse to adhere to
the diet. Most trials must therefore also be
analysed on the per-protocol basis.

The failure to complete a trial may have
serious consequences when the investigator
intends to complete a series of Latin or Graeco-
Latin squares so that order and carry-over
effects can be balanced and calculated. Often a
patient will default and remove the possibility
of an elegant and simple analysis (chapter 5).
The situation may be retrieved, however, by
substituting for missing values using multiple
regression or other techniques [219].

THE PROBLEM OF PATIENT NONCOMPLIANCE

Noncompliance is the failure to adhere to
therapeutic advice. Some noncompliant patients
may be willing to report this at interview or on
a self-administered questionnaire. Those who fail
to admit to noncompliance cannot be identified by

the physician but may be detected by pill count or by more objective measurements.

Failure to comply with therapeutic advice may have important consequences in randomised controlled trials. If the treatment regime is only adhered to by a proportion of subjects in the trial, the results for the patients will relate to the intention to treat rather than the effect of adhering to the particular intervention. If the patients in the trial are exceptionally compliant the effect of treatment may be large whereas if they are less compliant than expected the effect of treatment may be reduced, too few patients may be recruited to determine the effect, and any dose-response relationship may be underestimated. Lastly we must consider the misinterpretation of trial results that may result from noncompliance and the use (misuse) of noncompliant patients as a control group.

METHODS OF DETECTING NONCOMPLIANCE

Interview to detect noncompliance

An interview detects a proportion of noncompliant patients. When noncompliant patients admit to failing to adhere to their treatment, they are almost certainly not lying [220,221] and an interviewer easily identifies a proportion of such subjects. However, not all noncompliant persons are detected by a simple interview [222-224], and this method overstates the degree of adherence to a treatment regime. When analysing the results of a trial according to compliance as assessed from an interview we are unable to distinguish between patients who adhere to their treatment and noncompliant patients who refuse to admit to this fact [221].

Self-administered questionnaire to detect noncompliance

The completion of a self-administered questionnaire has the same advantages and disadvantages of the interviewer technique.

However, whereas in a small trial there is often ample opportunity for one interviewer to question the patients, if a large number of subjects are involved it is easier and cheaper to arrange for a standard questionnaire on compliance to be completed [225]. A self-administered psychological scale to detect obsessionality has been shown to produce a score inversely related to nonadherence. This may prove useful in detecting patients who are noncompliant [225]. However, such a technique has not yet been attempted in a clinical trial.

Physician's assessment of patients who do not admit to noncompliance

The physician's assessment has been shown to be useless when considering noncompliance with drugs [226]. The physician appears to have no way of predicting who will be noncompliant in the future nor who has been noncompliant during the course of a trial.

Pill count to detect noncompliance

For a pill count the patients are asked to bring their containers of tablets to each visit. The pretext for returning these is to determine whether or not a further prescription is required and the contents are counted out of sight of the patient. The observed number of pills left is compared with the expected number. The method is obviously open to manipulation by the patient and takes no account of the patient's giving other persons his treatment or otherwise destroying some of the tablets [227,228]. The use of a pill count may overstate the compliance of the patient.

Drug or metabolite concentrations in blood or urine

Compliance with a drug regime may be tested by measuring drug or metabolite concentrations in the blood or urine. However, this method does not necessarily estimate the day-to-day degree of noncompliance. For example, a patient may usually take his or her tablets but may not have taken them at the time of the blood test. Another patient may omit most of his medication but have taken the treatment just prior to the time the blood sample was taken. Patients may be more likely to take their medication on the day that they make a visit to the investigator, and blood samples at this time may underestimate noncompliance. The number of tablets taken cannot be determined as the pharmacokinetics of many drugs are complex and it is difficult in an individual to predict the concentrations of the drug or its metabolites that would be expected from a given consumption of the drug. If a drug is difficult to detect in the urine, the tablets may be labelled with a fluorescent dye and the urine examined for fluorescence.

Certain drugs have marked effects on blood constituents and these changes may be used to monitor noncompliance. For example, a thiazide diuretic may lower serum potassium and raise serum uric acid. A patient whose serum potassium is not lowered by treatment or in whom the serum uric acid does not change in the expected direction can be suspected of noncompliance [229]. Similarly, in a trial of antismoking advice the carboxyhaemoglobin level in the blood is an indication of whether the patient has stopped smoking or not.

Electronic monitoring methods

A pill bottle is available with a micro-processor in the cap to record every bottle opening [230-232]. The fact that the bottle is opened at a certain time and date does not prove that a tablet was taken, nor does the fact that the bottle was not opened prove noncompliance as tablets may have been transferred to a second container. Nevertheless Urquhart has identified 3 patterns of noncompliance; the 'toothbrush effect' when a noncompliant patient complies during the day or so prior to visiting the physician; the 'drug holiday' when several days elapse without treatment; and the steady decline in dosing which often starts a few days after the beginning of therapy [233]. It is claimed that electronic systems detect noncompliance when drug serum concentration and pill counts fail [232].

Objective measurements of compliance are to be preferred to indirect methods but may be expensive and relatively more difficult to organise than other measures of compliance.

THE CONSEQUENCES OF NONCOMPLIANCE

Owing to noncompliance, later trials of a new treatment may demonstrate a smaller effect than earlier trials

Early trials of new treatments (for example, a new pharmacological agent) will be performed on volunteers or selected patients in a laboratory setting. These subjects are likely to be more compliant than patients subsequently treated outside the research-institute environment. Assuming that noncompliant patients demonstrate a reduced drug effect, the average effect of a given dose of drug will be smaller when these patients are included. Later trials (for example, on outpatients) may therefore suggest a smaller pharmacological effect of the treatment owing to noncompliance.

This 'dilution effect' of noncompliance was well documented as early as 1962 [234]. Urquhart has calculated that the wilful discarding of many doses in a single moment prior to having a pill count may have over-estimated compliance in the Lipid Research Clinics - Coronary Primary Prevention Trial, so that cholestyramine at full dose may have lowered cholesterol by 25% not 19% [233]. This dumping of treatment has been called the 'parking lot effect' as this is where the drugs are found outside clinicians' offices.

The numbers required for a randomised controlled trial may be underestimated owing to noncompliance

When estimating the numbers required for a trial the investigator may employ the results obtained from a pilot or early trial where compliance is high [235]. The numbers required for the trial will be calculated on the basis of the effect shown in this trial; if the effect in the main trial is smaller due to noncompliance, then insufficient numbers may be recruited for the trial.

The dose-response relationship may be underestimated in noncompliant patients

If noncompliance is present, the effect of a particular drug dose will be underestimated and any estimate of the dose response will be incorrect. In addition, noncompliant patients may appear to take large doses of the drug with no adverse consequences. However, the difference between therapeutic and toxic doses may be small, and when later compliant patients are given the higher doses they may experience an adverse effect.

Noncompliance may lead to a smaller response in one of two equal treatments

In a randomised trial we compare one strategy with another. If one strategy proves more effective than another we may conclude that it is to be preferred, but we cannot necessarily conclude that the pharmacological agent given is more effective. For example, if two drugs are employed in a trial, one may cause side effects and another may not. It is possible that noncompliance will be greater with the drug producing adverse effects than with the other. The drug that induces noncompliance will have its pharmacological effect underestimated and the other drug may appear more pharmacologically active. However, the conclusion from the trial is correct in the sense that the strategy of prescribing the drugs leads to the observed results.

Feinstein provides another example when dietary advice is given [236]. The strategy of dietary advice is compared with no dietary advice and the end point of the trial is survival. If all the cases randomised to diet are compared to all randomised controls, then the results of the trial should be clear and any benefit from a strategy of suggesting the diet will be apparent.

However, it is possible that health-conscious people will comply with the dietary advice and persons who are not worried about their health will not. In this instance, noncompliant patients may continue to smoke and take no exercise and these habits may adversely influence their survival. If the patients who are noncompliant with dietary advice are omitted from the subsequent analysis of survival, then we will compare health-conscious people (who took the dietary advice) with the control patients who consist of both health-conscious and health-careless persons. It would not be surprising if such an analysis proved the diet to be beneficial. A similar example of antismoking advice was discussed in chapter 7. Again, such trials must be analysed on an intention-to-treat basis.

The deliberate use (misuse) of noncompliant patients as a control group

When a randomised controlled trial cannot be performed it is very difficult to estimate the gain that results from treatment. For example, when considering surgery for a resectable cancer a randomised controlled trial may not be possible but a control group, consisting of persons who refuse to have the operation, may be collected. However, the control group will consist of a very biased and unusual sample; such a study lies outside the scope of this book.

An overall view of noncompliance

Noncompliance by patients during a trial may lead to a misinterpretation of the results. However, the results of a trial that includes noncompliant patients may be more applicable to the effects of the treatment in the community. In a randomised controlled trial, procedures should be adopted to detect noncompliance and the results analysed in two ways: first including the noncompliant subjects, and the second omitting them. A proportion of noncompliant subjects may be detected by simple and inexpensive interview methods, and urine or blood tests may provide more objective measures of compliance on a single occasion. Noncompliance may also distort any dose-response relationship and lead to inadequate numbers being recruited to a trial.

NONADHERENCE TO THE PROTOCOL BY INVESTIGATORS

Chapter 8 considered the importance of drawing up a detailed protocol and it is important that all the participants agree to adhere exactly to the protocol. Occasionally the investigator may have to deviate from the protocol in the interest of an individual patient. Such an event will usually lead to the termination of the trial for that particular patient. More important is consistent failure of the investigator to adhere to the protocol. This may not be readily admitted and has to be determined indirectly. Nonadherence can be accidental or intentional and involves four most important areas: deliberate major fraud, admission to the trial of patients who should not be admitted, breaking a double-blind code, and the prescription of additional treatment that is not allowed in the trial protocol.

Deliberate major fraud

In 1975, Dr JPS, a general practitioner in the UK, was removed from the Register of doctors allowed to practice in the UK [237]. He had contracted with Bayer UK Ltd to perform a double-blind cross-over trial comparing an anti-hypertensive drug with placebo. For each patient up to 100 he received £10. He submitted 101 reports signed by seven different doctors. However, the results showed that the active drug had *"a uniform and consistent effect which was markedly different from test results from other sources"*. It transpired that the other 6 doctors did not take part in the trial and the trial data were assumed to be fraudulent.

In 1978, Dr MJS was the principal investigator of a 40 strong research team receiving almost $1 million from the National Cancer Institute. His team were collaborating in a multicentre drug trial organised by the Eastern Co-operative Oncology Group and falsified data were provided [238]. Dr MJS later sued five former members of his team.

In the UK in 1986, Reckitt and Colman arranged a multicentre trial of an anti-depressant drug and Dr VAS, a psychiatrist in Durham was recruited. Anxiety about his data arose when he submitted reports on all fifteen patients at once and these forms were too neat [238-240]. The Clinical Research Associate (CRA) tried to check the source documents and was unable to do so. Dr VAS reported that the trial had been performed by his registrar. When finally tracked down, because she had left and he could not remember her name, she denied any involvement. Dr VAS' name was erased from the medical Register but this had an unfortunate consequence. The local medical community was *"horrified and incredulous. How dare .. a pharmaceutical company do this to one of their highly respected colleagues?"* [240]. The immediate effect was that the company was banned from access to the Postgraduate Medical Centre. This ruling was quickly rescinded but Wells has reported *"No pharmaceutical company was going to go through this again if it were to be jeopardised as a result"* [240].

Wells has reported on four other cases of fraud, detected by: patients all arriving on the same day and being very similar; patients not receiving any concomitant medication; not having any adverse events; data varying considerably according to source (peak expiratory flow rates differing markedly when recorded by patient at home and by the doctor in his office); inconsistent data (for example, distance travelled from home to surgery); patient completed information all in the same handwriting; patient's handwriting changing during the course of a trial; unused tablets in strip and blister packs returned in the same condition for every patient (often in pristine condition); dates incorrect or impossible; no errors or changes of mind in self-administered questionnaires; no withdrawals from trial; failure to allow verification of data by a third party; requests for copies of forms back so that records can be 'got ready'; and unbelievable data. Unbelievable data include one trial of 19 patients where 8, 7 and 4 patients had identical ECGs, where findings were consistent beyond statistical belief, and where an asthmatic woman of 60 walked at 7.2 miles/hour.

It is hoped that pharmaceutical companies will not now be deterred from pursuing fraudulent investigators and that they have appropriate screening methods for detecting fraudulent results. Good Clinical Practice (GCP) guidelines have now been introduced to avoid these problems and are discussed later in this chapter.

Admission to the trial of patients who do not fulfil the admission criteria

The coordinating office or the person randomising the patients should confirm whether or not the patients are eligible for the trial. For example, a trial may admit persons with a serum uric acid within a certain range and the coordinating centre must refuse to randomise a patient who does not have a uric acid within this range. An unintentional failure to adhere to the protocol was observed in the University Group Diabetes Program trial where several patients were admitted to the trial without the strict criteria of diabetes mellitus [63]. This problem added to the considerable controversy about this trial, which is discussed in chapter 19.

Patients are occasionally admitted to a trial owing to administrative errors - for example, when a patient is admitted before an investigation result is available. The patient may have fulfilled the other entry criteria to the trial but when the result becomes known it may exclude the patient. Administrative errors may also occur when noncompliant patients are to be excluded by pill counts. A patient may repeatedly forget to return with his tablets and compliance with medication cannot be confirmed.

Breaking a double-blind code

An investigator may deliberately break a double-blind code without good reason by examining the details of treatment secured in a sealed envelope. This is unlikely to occur, but he may guess the identity of the treatment from other information. For example, in trials of beta-adrenoceptor blocking drugs the investigator who is prescribing the treatment should not measure the pulse rate as he may detect the marked slowing of the pulse that occurs with this treatment. For this reason one investigator should prescribe and a second investigator assess the results of treatment.

Treatment during the course of a trial

During a trial of drug or dietary treatment other treatments may interfere with the results of the trial and may be prescribed either by accident or design. When comparing active antihypertensive medication with placebo a protocol may not allow the administration of other pressure-lowering drugs. If a patient develops angina, he may require a beta-adrenoceptor blocking drug and have to be withdrawn from the trial as this treatment also lowers blood pressure. The investigator must adhere to the protocol and withdraw the patient.

Conclusions on deviations from protocol

Quality control is critical throughout the trial and should detect deviations from the protocol. With GCP Guidelines poor or fraudulent information should be detected and these guidelines are now discussed.

GOOD CLINICAL PRACTICE GUIDELINES

Good Clinical Practice (GCP) in the USA and EU is a set of management procedures for clinical trials intended to prevent fraud and errors and to protect patients' or volunteers' rights. The background to the introduction of GCP has been described by Allen *et al* in their useful book entitled 'Good Clinical Practice in Europe - Investigators Handbook' [241]. GCP was first introduced in the USA. The Food and Drug Administration (FDA) proposed regulating the actions of investigators and sponsors in 1977 following trials on prisoners, mentally defective children and poor and badly educated black men. In addition it became apparent that fraudulent reporting by investigators did occur. The regulations were introduced and investigators who seriously infringed these were not allowed to perform trials on new drugs, and the sponsoring companies who did not comply with the regulations failed to have their drugs approved [241]. For about ten years studies were performed in the USA to GCP standards but not necessarily in the EU. In July 1990, the European GCP guidelines were approved and became effective in July 1991. The American GCP requirements involve more information being given prior to informed consent and the American Ethics Committees are involved throughout the trial. Nevertheless, the requirements are basically similar and a short summary is given below. The reader is directed to references 241-245 for further reading and a journal started in April 1994 that is solely concerned with this topic - the 'Good Clinical Practice Journal'.

SUMMARY OF GCP GUIDELINES

This structure is based on the book by Allen *et al* [241].

1. *Protection of the Individual (Informed consent and Ethics Committee review)*
The subject must give his or her signature on a written consent form. The form must state that research is being conducted, and why and what exactly will happen. The subject needs to know the procedures, known risks, discomfort and duration of the study. Potential benefits (if any) should be mentioned together with alternative treatments available, the fact that an outside body may inspect their medical records, arrangements for compensation, and a contact number for information. Most importantly it must be clear that participation is voluntary and the patients may withdraw at any time without impairment of usual treatment.

The local ethical committee, or Institutional Review Board (IRB) in the USA, must approve the protocol. In particular the committee will ensure that informed consent is provided, and will also review any protocol amendments. Many committees also review the outcome of trials, require re-consideration of approval after a interval of say, 4 years, and need to be informed of any other ethical committee that has rejected the protocol. In the USA the board is responsible for a continuing review of the research. The IRB has to have at least five members, of varying backgrounds, not all men or all women, must include a lay member, include persons outside the institution and exclude persons with a conflict of interest.

2.   *Standard Operating Procedures (SOPs)*
A standard operating procedure is a procedure covered by a detailed manual of operations. A SOP informs all personnel involved exactly what should be done and when. The SOPs cover

i)   responsibilities of staff and reporting structure.
ii)   a definition of required meetings, eg. at start and end of study, monitoring visits
iii)   quality assurance (see below)
iv)   monitoring visits
v)   handling of study drug
vi)   adverse event reporting (see below)
vii)   techniques of measurement eg. blood pressure, cognitive function
viii) transport of specimens
ix)   financial arrangements, including payment for travelling expenses and investigations

3.   *Documenting and archiving*
This includes the provision of a Case Record Form (CRF) which must be filled in accurately, alterations made without obscuring the previous entry, and original record and alterations signed by the investigator. CRFs have to be retained for at least 15 years after the end of the trial. These records may be returned, sealed, to the sponsoring company [241].

4. *Source Data Validation (SDV)*

All data must be available for audit by the sponsor or the drug regulatory authority. In the USA, data validation takes place for patient identity, consent form, eligibility for inclusion, medical history, outcome measures, adverse event records, and laboratory and other results. The audits are performed by monitors from the sponsoring company, independent persons, and in the USA and France, the drug regulatory authorities.

5. *Adverse Drug Reactions (ADRs)*

The investigator must report serious suspected adverse drug reactions to the sponsor without delay. The sponsor has a responsibility to report these to the drug regulatory authority immediately. Possible serious adverse drug reactions include any reason for being admitted to hospital and therefore constitute a common problem in trials in the elderly. The regulatory authorities differ in their requirements, especially in the time that may elapse between the event and their receiving a report. The authorities also require to be informed about minor ADRs but these requirements vary widely.

CONCLUSIONS

This chapter examined some problems that will be encountered during the conduct of a trial. The investigator should first consider performing a pilot trial or at least starting the trial in a pilot fashion. The advantages and disadvantages of a run-in period were discussed together with the problems of reducing dropout, detecting patient noncompliance, and the failure of investigators to adhere to the protocol. To avoid the collection of fraudulent and erroneous data, Good Clinical Practice Guidelines have been introduced and these are discussed.

The problems in execution of a trial should be presented in the report of the trial and information should be given both on procedures that worked well and those that did not. The reader must understand that errors will occur in the performance of trials and that the results cannot be discounted on the basis of minor deviations from a protocol. Details of such errors should be reported together with the number of patients considered for the trial, entered in the study, and withdrawn.

12. STOPPING RULES FOR TRIALS

The duration of treatment for the individual patient is specified at the design stage as is the total number of subjects to enter the trial. In the normal course of events the trial will last as long as it takes to enter the required number of patients and for the last patient to complete the study. However, the trial may be abandoned early if adverse effects of treatment are observed, if a benefit is demonstrated at the predetermined level of significance, or if it is apparent that the required number of patients will never be recruited or that the response to treatment is not as expected.

From the ethical point of view it is important to minimise the number of subjects receiving the inferior treatment. This will be achieved by stopping the trial when an 'almost certain' advantage has been detected for one of the randomised groups. The ethical aspects are discussed in chapter 3, pages 41-43.

Decision rules for stopping the trial must be agreed at the design stage and in a long-term trial a data monitoring committee or Data and Safety Monitoring Board (DSMB), must examine the results of the trial at given points in time. The decision to stop must be made by an independent committee, such as the Advisory Board discussed in chapter 8.

The decision rules for stopping the trial will be discussed together with the disadvantages of stopping too early or continuing too long. In addition we shall consider when the results of one trial should stop other trials that are underway or planned to start.

Decision rules for stopping the trial

Table 12-1 summarises the general format of stopping rules. The table assumes that interim analyses are performed at predetermined intervals

Table 12-1. Decision rules for stopping a long-term trial with morbidity or mortality as an end point.

RULE 1

A statistically significant increase in a serious adverse effect is observed in an actively treated group. Minimum adverse effect rate exceeded.

RULE 2

A statistically significant decrease in morbidity or mortality is observed. Minimum benefit exceeded.
plus
A reduction in total morbidity and mortality is observed commensurate with there being no transfer of morbidity or mortality from one cause (the end point of the trial) to another.

RULE 3

The predetermined number of patients has been admitted to the trial and followed for a given length of time.

RULE 4

The number of patients recruited will not be adequate or the effect of treatment is not as great as expected.

RULE 5

Evidence accrues from other sources that make it unethical to proceed.

---

and that the level of significance is adjusted for repeated looks (chapter 6).

Rule 1 states that the trial must be terminated if a statistically significant and biologically important adverse effect of treatment is demonstrated. Biological importance will be determined from the severity of the adverse effect and its frequency. Rule 2 states that a trial must be designed to terminate when any one important benefit is demonstrated provided the effect is compatible with an overall benefit to the patients. A reduction in total mortality and morbidity must be apparent, although probably not statistically significant. A treatment may reduce, say, myocardial infarction or stroke yet produce an excess of other serious morbid or mortal events. An antihypertensive drug may reduce stroke events but produce episodes of hypotension and an excess of injuries due to falls and other accidents. When considering terminating such a trial because of a reduction in stroke events, we must be certain that there has been no comparable excess of morbidity from other causes. The trial should only terminate when the benefit : risk ratio is acceptable.

It has been suggested that in trials of secondary prevention in myocardial infarction both mortality from myocardial infarction and total mortality must be reduced with statistical significance to stop the trial. However, such a trial may arrive at a statistically significant reduction in myocardial infarction events before total mortality is significantly reduced. It would be unethical to continue the trial if total mortality is also reduced by a commensurate but non significant amount.

Rule 3 states that a trial should be terminated when the intended number of patients has been recruited and followed, even when there is a negative result but with the power as specified in the trial design. Rule 4 states that a trial may be aborted if there is no hope of recruiting the required numbers for a given amount of time or money. The trial may also be stopped if the treatment is not producing the effect anticipated in the design. For example, it may be intended to assess the effect on myocardial infarction mortality of lowering serum cholesterol by 1 mmol/l. A diet may be chosen for this purpose but prove to lower serum cholesterol by only half the expected amount. In this event the intended reduction in mortality will be unlikely to be observed and the trial may have to be abandoned. Rule 5 considers the situation where other similar trials are published with clear results of substantial benefit (or harm) and the investigators decide that it is unethical to continue.

Problems in stopping a trial early

i) *Practical difficulties*
Many logistical problems arise when a trial is stopped [246]. For example, the End Point Committee may not have validated all the

preliminary events. In the Systolic Hypertension in the Elderly Program (SHEP) trial the median time between the occurrence of a stroke and entry to the data base was 7 months [247]. Thus the final analysis will not exactly coincide with the result when the trial is terminated. This situation is exaggerated by the need to see the patients when the trial is terminated in order to stop the treatment. This process is expected to reveal many more events. The Data and Safety Monitoring Board (DSMB) overseeing the SHEP trial was also worried that if it stopped the trial early the results would have to be communicated quickly and they preferred to release the data with *"orderly scientific review"*. The DSMB therefore allowed the trial to proceed a few months later than necessary and at the scheduled end of the trial [247].

ii)   *Repeated looks at the data*
The necessity for repeated looks at the data increases the overall type 1 error (chapter 6). This has to be allowed for in the stopping rules. Moreover, most would not like the trial stopped for a given constant P value but would prefer that the trial be stopped early only for a very high probability value and stopped later for more modest results.

iii) *Timing of repeated looks*
The DSMB should not examine the data at fixed time intervals as initially there may be too few events to expect any conclusion. The DSMB should examine the data at intervals determined by the number of events.

iv)   *Should there be formal sequential boundaries indicating when to stop the trial?*
Although this was rejected in the SHEP trial because of *"the strong desire of the investigators to learn about the long-term effect of antihypertensive medication in the elderly"* [247] the stopping rules are more open to manipulation if they have not been finalised. Most would prefer a guideline stopping rule which could, of course, be ignored if the benefits were accompanied by an excess of

risks. In other words if the benefit : risk ratio is not conclusive the trial should continue.

v)   *The problems of stopping the trial too soon*
The result of a trial may be accepted if it agrees with the preconceived notions of the medical community and rejected if it is unexpected (chapter 19). With a surprise outcome it is important to achieve a high level of significance, preferably less than one percent. The pressure to terminate a trial increases as a significant positive result approaches and the temptation must be resisted until the desired level of significance is achieved. One large trial was continued in an attempt to achieve a higher level of significance and will be discussed in chapter 19. If a trial is intended to prove the efficacy of a treatment that has appeared beneficial in observational studies, then a lower level of significance may suffice.

A leading article in the Lancet highlighted the problem of stopping a trial too early [248]. The ACTG019 trial of zidovudine for symptomless HIV infection was stopped with a positive result after a mean follow-up of only one year. The result of this trial is now questioned and the leading article asks *"how should the result after early stopping be interpreted?"*.

vi)   *Should only a P value stop the trial?*
A P value should not be sufficient to stop a trial. The benefit : risk ratio has to be evaluated and also the <u>magnitude</u> of the effect must be taken into account. Meier has formalised some general stopping rules for a clinical trial [104]. He has defined the maximum acceptable difference (MD) as the largest true difference between treatments that a subject in the trial should be expected to accept and yet continue in the trial. He also defined the least interesting difference (LD) as a true difference, although not terminating the trial, of sufficient magnitude that when exceeded the difference would *"be enough to justify a decision in favour of the winning therapy"*. Figure 12-1 illustrates the possible use of such rules. At the first analyses the MD is

238

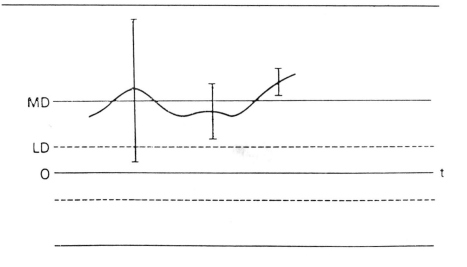

MD-Maximum acceptable difference
LD-Least interesting difference

Figure 12-1.   The difference between two treatments and confidence interval is plotted against time. The trial will be stopped when the MD is exceeded with statistical significance. Three analyses are performed: at the first, the maximum acceptable difference is exceeded but not with any confidence; at the second analysis, the result lies between the LD and the maximum acceptable difference; and at the third analysis, MD is exceeded with statistical confidence and the trial is terminated. Adapted from Meier, Clin Pharmacol Ther 1979;25:649-650, with permission.

exceeded but not with any confidence. At the second analysis the result lies between MD and LD, but at the third analysis the result exceeds MD with statistical confidence and the trial is terminated.

vii) *The disadvantages of stopping the trial too late*
When a trial is terminated and a very high level of significance for a large effect has been achieved, the question arises whether the trial could have been terminated earlier and the benefit of active treatment given to the control group at that time. For example, in a trial of secondary prevention of myocardial infarction there were 98 deaths on the active treatment

(timolol) and 152 deaths on placebo ($P = 0.0003$) [218]. It is theoretically possible that some of the placebo deaths could have been prevented if the trial had been terminated earlier when $P = 0.01$. However, the review committee may have only examined the data when $P > 0.01$ and later when $P = 0.0003$ and therefore had no opportunity to stop the trial at an earlier stage.

In the ISIS-2 trial the primary comparison had a P value less than 0.00000001 [77,250]. In this trial the steering committee required both proof beyond reasonable doubt and evidence that would materially change patient management by clinicians. The investigators were informed of the positive result in a subgroup [79] but the trial continued. The investigators were allowed to continue randomisation if they were uncertain about whether or not the treatment was indicated in an individual patient. They continued to enter patients to establish the benefit in different subgroups. In the ISIS-3 trial this strategy was formalised in the protocol, [248,251].

The problem of stopping a trial too late may be increased when treatment has an adverse effect [252].

Stopping a trial because of accumulating information from other trials

When the ACTG019 trial stopped the Concorde trial went ahead with randomisation and follow-up. The preliminary negative results for zidovudine in the Concorde trial was the reason why the results of the earlier trial were questioned [248,253]. It is fortunate that the results of a single trial were not considered conclusive and literally the 'last word' on the subject. Usually several trials address the same issue but sometimes one large trial will convince the enthusiasts that a particular treatment is obligatory and they will suggest that further trials are unnecessary and unethical. One example is the treatment of colon cancer with adjuvant 5-fluorouracil and levamisole after surgery where trials continue despite one positive trial [248,254]. A second example is given by the SHEP trial which reported during the progress of similar trials to treat isolated systolic hypertension in the elderly, the SYST-EUR and SYST-CHINA trials [183,255]. These trials continued for the following reasons.

i)   the benefit : risk ratio in the SHEP trial had not been clearly established as adverse effects of treatment were observed [184,185,256].

ii) a minimal level of benefit had not been exceeded [189].

iii) the results in the USA may not be generalisable to other populations [186]. The results of SHEP may be expected to differ from the Results of the European and Chinese trials as these trials employ a different treatment regimen [183,255]. However it may be argued that these trials should stop and treatment changed to that employed in the SHEP trial.

In connection with generalisability, trials of mammographic screening continue in the UK despite benefits revealed in a Swedish study. It was considered that the uptake rates, specificity and sensitivity could not be generalised to the UK [187-189].

iv) given the SHEP results what is the probability that a second trial will detect a benefit of at least 15% with a probability of 1%? For stroke events this is 0.36, in other words a 64% chance of not finding this size of benefit. Similarly this level of benefit would be found with a probability of 0.20 and 0.06 for stroke mortality and cardiovascular mortality respectively [189].

The SYST-EUR and SYST-CHINA trials continued owing to the above uncertainties and the need to establish benefit beyond doubt before large numbers of elderly subjects were treated in different countries.

DIFFERENT METHODS OF PROVIDING STOPPING GUIDELINES

1.   The O'Brien, Fleming method [257]

This was employed in the ß-Blocker Heart Attack Trial (BHAT, [258,259]) and the West of Scotland

Coronary Prevention Study [260]. The method requires that early stopping of a trial requires very small P values, whereas later values for stopping are closer to conventional levels of significance. Pocock argued that this is justifiable as the medical community is sceptical of the results of small trials, early large benefits from treatment are unlikely and stopping early with higher P values reduces the power to detect other effects [250]. To these could be added that increased numbers at later stages will reduce the confidence interval of the result.

An example of stopping guidelines has been provided by the West of Scotland Coronary Prevention Study [250,260], five analyses will be performed with the following guideline P values; analysis 1, $P < 0.00000001$; 2, $P < 0.0001$; 3, $P < 0.001$; 4, $P < 0.004$; and the final analyses, $P < 0.009$. This gives an overall two-sided significance level of 0.01.

The O'Brien, Fleming method is also employed in the SYST-EUR trial with four interim analyses after 50, 100, 150 and 200 stroke events and a final analysis [183]. Asymmetric 2 sided boundaries have been calculated giving an overall P value of 5% for an adverse effect and 1% for a benefit. It is important to note that the protocol states:

"Although the interim analyses will provide major guidance in the decision to stop the trial prematurely, rigid stopping rules will not be imposed, as they cannot reflect all possible outcomes that might arise during its course."

When the O'Brien and Fleming rule was applied in a trial of chemotherapy plus radiation with radiation alone in lung cancer [261], the trial was stopped early with $P = 0.007$ in favour of the joint treatment. As chemotherapy is toxic and expensive it was very important that the benefit : risk ratio was exactly quantified and stopping this trial early has been criticised [262]. There is great concern that stopping early on a 'random high' will lead to a false positive result.

2. Monitoring conditional power

This is essentially the probability that the trial would eventually lead to the same conclusion, even if treatment conferred no further benefit. This method was employed in the SHEP trial [247]. The Drug and Safety Monitoring Board (DSMB) met between December 1985 and December 1990 and voted at meetings whether or not to stop the trial. They voted 'no'in December 1987 when total mortality was lower in the actively treated group and the conditional power was only 15%. The 'reduction' in total mortality observed at that time (Z = 2.4) declined to Z = 0.5 later. In July 1990 the conditional power was 87% (Z = 3.5) for total strokes but the trial was not stopped because of the logistical problems mentioned above.

3. Constant critical value (low P) during the trial with a large critical value (higher P) at the end

The Peto *et al* rule [132] specified a constant low *P* value for stopping early, for example *P* < 0.001. This method is being employed in the European Myocardial Infarction Amiodarone Trial (EMIAT) and does allow flexibility in the number of interim analyses to be performed [250]. However the trial is more likely to stop early with this guideline.

CONCLUSIONS

It is an ethical necessity to monitor progress in a long term trial to ensure that future participants do not suffer from excessive adverse effects or from not receiving an effective treatment. Trials have been stopped early from an adverse effect, for example the Cardiac Arrhythmia Suppression Trial (CAST, [263]); more commonly from the demonstration of benefit; and when there appeared to be no point in proceeding. The latter may occur when the recruitment rate is impossibly low, when outcome events are not being observed or when the benefit : risk ratio will never justify treatment. This occurred with a trial of neutron therapy compared to conventional radiotherapy for pelvic cancer [252]. To change clinical practice a relative risk of 0.7 was thought necessary. When the relative risk was 1.52 (95% CI 0.91, 2.50) the investigators stopped the trial as it was considered pointless to go on.

Crucial to the consideration of guidelines is the existence of an independent Data Monitoring Committee (DMC) or Data and Safety Monitoring Board (DSMB) to give the committee its American title. The committee is responsible for advising on whether or not the trial should continue. However the DMC or DSMB has to look not just at the guidelines, certainly not only at the P value, but at the overall benefit : risk ratio (chapter 17). Thus in the SHEP trial the DSMB examined the baseline results in the two groups; adherence to the protocol by staff and patients; accuracy and completeness of the database (in particular had the vast majority of patients been adequately followed-up?); any evidence for bias in ascertainment of outcome; results in different subgroups and centres; benefits other than stroke reduction; and level of risks [247]. It is also important that the committee is assured that no data have been fabricated. If the trial is run under Good Clinical Practice Guidelines there should be no anxiety over this. Lastly the committee should consist of experts who are not entering subjects into the trial so that their clinical management cannot be influenced by an awareness of the results. The trial statistician must be part of the committee but should not be a voting member. The trial sponsors should not attend the meetings and should not be informed of the results. Sponsors, especially from the pharmaceutical industry, have vested interests in both early publication of positive results and aborting a negative trial to save further expenditure.

13.   THE VARIABILITY AND VALIDITY OF RESULTS

Even when the results of a trial are free of bias, the measurements on which they are based are still subject to random variability. Often the variability can be estimated from its repeatability and reduced by quality control. The measurements may vary more between subjects than within subjects and this variability can be reduced by employing the within subject cross-over trial design. It must also be remembered that a measurement may be very repeatable but either not provide the correct result (low accuracy) or measure something other than that intended (low validity).

ACCURACY

Accuracy is measured by the difference between the measurement and the known true result. For example an exact amount of glucose could be made into a solution and this concentration compared with the results of a machine analysis. In this example the 'truth' is known but often we have to compare one machine with another. One machine has to be considered to be the gold standard against which the second machine is compared. Accuracy is usually given by the average difference between the measurement and the gold standard. If the differences include the sign (+ or -) then the tendency to over-read or under-read (the bias) is given. If the differences are averaged irrespective of sign a better description of accuracy is given.

REPEATABILITY

Repeatability is the level of agreement between replicate measurements. The level of agreement must be determined for measurements made in the

same subject in order to exclude differences
between persons. The variability may result from
observer error, machine error and, when
repeatability is estimated at widely spaced
intervals, from alterations in the true
measurement over the time of estimation.

Repeatability of a continuously distributed variable

The repeatability of a continuously distributed
variable is estimated by the standard deviation
of repeated measurements. The standard deviation
of repeated measurements has been termed the
precision of the measurement [264]. When the
deviation is high it will be less precise. The
standard deviation is often divided by the mean
and expressed as a percentage known as the
coefficient of variation.

Obviously, it is important to make very
repeatable measurements during the course of a
trial and great difficulties have arisen when
this has not been the case. The World Health
Organisation and the Centre for Disease Control,
Atlanta, have been involved in assessing the
repeatability of serum cholesterol measurements
around the world [264]. Cooper has reported these
results and found that, although the standard
deviation of the various methods was
satisfactory, some methods give high readings of
cholesterol and poor accuracy. The two methods
causing the most problems were the direct method
of cholesterol estimation and the chloride
method. It is worth emphasising that if a
laboratory changes its method of analysing
cholesterol or any other substance during the
course of a trial, the measurements may be
altered with serious consequences. The new method
may differ in precision or accuracy when compared
with the old method.

Figure 13-1 illustrates the situation when
a biochemical measurement on blood is checked
three times during a clinical trial. Each subject
has blood taken on three occasions and the blood
is split into three samples. Biochemical
estimations on the samples give the results
designated X. The results for the first subject

	OCCASION SPECIMEN TAKEN									
	1			2			3			Average
	Split sample			Split sample			Split sample			
	1	2	3	1	2	3	1	2	3	
Subject 1	X_{111}	X_{112}	X_{113}	X_{121}	X_{122}	X_{123}	X_{131}	X_{132}	X_{133}	$\bar{X}_{1..}$
Subject 2	X_{211}	X_{212}	X_{213}	X_{221}	.	.	.	.	.	$\bar{X}_{2..}$
Subject 3	X_{311}	X_{312}	X_{313}	.	.	.	.	.	.	$\bar{X}_{3..}$
	.	.	.	.	.	.	.	.	.	.
	.	.	.	.	.	.	.	.	.	.
	.	.	.	.	.	.	.	.	.	.
Subject n	X_{n11}	X_{n12}	.	.	.	.	.	.	.	$\bar{X}_{n..}$
Overall mean		$\bar{X}_{.1.}$			$\bar{X}_{.2.}$			$\bar{X}_{.3.}$		$\bar{X}_{...}$

Figure 13-1. The assessment of repeatability during a clinical trial (see text). The data provide information on the precision and accuracy of measurements on each occasion.

have a suffix 1 (X_1); his or her results for the first occasion are designated X_{11} and results for the first sample on that occasion X_{111}. The subjects are numbered 1,2,3........n in the first suffix which we shall call i; occasions 1,2 and 3 are given the second suffix; and samples 1,2, and 3 the third suffix(s). The average for a suffix in the figure is given by a point. The average result for all measurements is given by $\bar{X}...$, the average for subject 1, $\bar{X}_1..$, and for occasion 1, $\bar{X}._1.$, etc.

The repeatability on one occasion
The within-specimen variance for subject 1, occasion 1, is given by

$$\sum_{s=1,r} \frac{(X_{11s}-\bar{\bar{X}}_{11.})^2}{r-1}$$

where $\sum_{s=1,r}$ is an instruction summing the squared differences between a sample (s) and the sample mean for s=1 to r, where r is the number of samples, which in this example is 3.

For all subjects and occasion 1 the precision of the measurement

$$= \sqrt{\left(\sum_{i=1,n} \sum_{s=1,r} \frac{(X_{i1s}-\bar{\bar{X}}_{i1.})^2}{r-1}\right)\bigg/ n}$$

where the squared differences are summed from i=1 to n subjects and s=1 to r (=3) samples.

In the same manner the repeatability of the measurement should be estimated for occasions 2 and 3 and compared with the initial measurement of repeatability. A worked example is given in the Appendix to this chapter to compare the blood pressure measurements on two prototype samples of a new kind of sphygmomanometer with one mercury sphygmomanometer. The prototypes did not prove to be acceptable.

The accuracy of the measurement can be gauged by comparing the mean for occasion 1 ($\bar{X}_{.1.}$) with the means for the other two occasions or, similarly, the average result for a standard machine with the average for the machine under investigation.

Another method can be employed in a long-term trial to identify any drift in the results. For each measurement during the trial the difference is calculated between this figure and the initial average results. The positive differences should balance the negative ones, giving a cumulative sum of differences near to zero. The cumulative sum of differences is calculated during the course of the trial and plotted against time. A deviation from zero increasing with time will indicate a change in accuracy. The graph has been termed a *CUSUM plot* and has been employed in monitoring laboratory performance.

Repeatability of qualitative data

Qualitative data are not continuously distributed and include diagnoses, symptoms, and clinical states given a discrete value (for example, zero when the condition is absent and one when present). Measuring the repeatability of qualitative data involves comparing two results (for example, when radiologists report a diagnosis). To estimate between-observer repeatability two radiologists may be asked to examine the same radiographs and report their opinions. If one observer is given the same films to examine on two occasions then within-observer repeatability can be estimated.

Figure 13-2 illustrates how the data may be presented, the columns giving the opinions of the first observer and the rows the results from the

FIRST OBSERVER

		Diagnosis +	Diagnosis -	Row total
	Diagnosis +	a	b	a+b
SECOND OBSERVER	Diagnosis -	c	d	c+d
	Column total	a+c	b+d	a+b +c+d

Figure 13-2. How the data may be presented to demonstrate the agreement between two radiologists examining the same films; a, b, c, and d are the number of pairs where they agree or disagree.

second observer. The letter "a" represents the number of positive responses in both assessments; "d" represents the number of negative responses on both occasions; "b" assessments were negative with the first observer but positive with the second observer; and "c" assessments were positive with the first observer and negative with the second. Three measures of agreement may be employed.

i)   *Kappa statistic* [265-267]
This statistic is to be preferred as it takes into account the agreement to be expected by chance. Basically this measurement is the ratio of actual agreement (over and above chance) to the maximum possible agreement (over and above chance). The expected agreement from chance (Ie)

$$\left[\frac{\frac{(a+b)\ (a+c)}{(a+b+c+d)} + \frac{(b+d)\ (c+d)}{(a+b+c+d)}}{(a+b+c+d)} \right] \quad ..(13-1)$$

is subtracted from the actual agreement (Io)

$$\left[\frac{(a+d)}{(a+b+c+d)} \right] \quad ....(13-2)$$

to give the actual agreement over and above chance. The maximum possible agreement over and above chance is 1 less the expected agreement from chance (Ie). The Kappa statistic (%) is the excess beyond chance divided by the maximum possible excess

$$\hat{K} = \frac{Io - Ie}{1 - Ie} \times 100 \quad ....(13-3)$$

Thus the example in figure 13-3 gives a Kappa statistic of 37.5%. The statistical significance of this measure can be calculated [268].
    Values of Kappa above 75% may be taken to represent excellent agreement beyond chance, values between 40% and 75% represent fair to good agreement and values below 40%, poor agreement beyond chance [268,269]. The agreement in Figure

FIRST ASSESSMENT

20% Prevalence		+	−	TOTAL
S E +		10	10	20
C O − N		10	70	80
D TOTAL		20	80	100

$$R_2 = \frac{80}{100} = 80\%$$

Figure 13-3. One measure of agreement,
$R_2 = (a+d)/(a+b+c+d)$, is 80%.

13-3 is therefore not much better by chance alone. If agreement is to be tested for different grades of a condition, for example severe, moderate, mild and absent, then a weighted Kappa statistic may be calculated [268].

ii) *The Mean Pair Agreement Index*

$$R_2 = \frac{a + d}{a + b + c + d}$$

This index is most easily understood but does not allow for chance agreement. Figure 13-3 shows the result of calculating R_2 in a condition with a 20 percent prevalence in both assessments and an equal division between the positive and negative cases. The index (R_2) was 80 percent.

A third measure of repeatability has been published.

iii) *Measure 3*

$$R_3 = \frac{1}{4} \left[\frac{a}{a + b} + \frac{a}{a + c} + \frac{d}{c + d} + \frac{d}{b + d} \right]$$

This index can be calculated as long as the sum of a row or column is not zero [270]. In the above example $R_2 = 69\%$.

Repeatability indices can be used to compare diagnoses made by different observers, the same observers on different occasions, or by different means (for example, electrocardiographic evidence of myocardial infarction compared with enzymatic data). Indices have been used to compare the responses of patients to self-administered questionnaires on two different occasions. In one study the repeatability of single questions was judged satisfactory but if a condition relied on more than one question the agreement was sometimes less than satisfactory [271]. For example, the subjects were asked if they had headaches and whether these headaches occurred on waking. The repeatability index for the pair of questions eliciting the response, waking headache, was low. Eliciting data on symptoms is discussed further in chapter 15.

The relationship between bias and repeatability

If a measurement is very variable, with a low repeatability, replicating measurements and taking the average may overcome the problem. In drug trials on hypertensive patients, repeated blood pressure measurements may be taken before and after treatment. By including sufficient observations and enough patients in the trial (see chapter 6) the results may be acceptable as a large number of patients and readings will give an accurate mean with low standard error. With smaller numbers, poor repeatability due to random error may obscure the beneficial effects of a drug or relationships between measured variables, but *bias* may produce a positive result that is incorrect. In within-patient cross-over trials the order effect of giving the drug represents a possible bias. An example was provided in chapter

7 where a subject may become more used to a blood pressure measurement between the first and second stages of the trial and the blood pressure may fall, not due to therapy but due to an order effect. If two drugs are being compared one can be started first in a random half of patients and the other first in the remaining subjects. If the order effect is not balanced out by this strategy the results will be biased in favour of the second treatment. Poor repeatability may lead to an answer not being obtained but bias may lead to an incorrect but positive result in favour of one treatment.

The effect of regression to the mean on repeatability

Regression to the mean was defined by Francis Galton in 1886 as *"each peculiarity in a man is shared by his relatives, but on the average to a lesser degree."* His definition meant that a very tall man was likely to have shorter brothers and shorter sons and a very short man taller brothers and taller sons [272]. The concept has been extended to describe the change in a variable through time that results from selection at the start. Many examples of regression to the mean exist and a good example is provided when blood pressures are measured. Subjects with pressures initially higher than the mean tend to have lower pressures on the second occasion and those with starting lower pressures than the mean tend to be higher on the second occasion. Therefore, if you select a group of hypertensive patients on the basis of their high blood pressures, their mean measurement will certainly be lower on the second occasion owing simply to this regression-to-the-mean phenomenon. This fall in pressure and loss of repeatability should occur both in an actively treated and a control group and the provision of the control group will prevent the effect being ascribed to active treatment. Regression to the mean can also be limited to some extent by only including patients with a repeatedly high pressure, such patients presumably having a smaller tendency to drop to lower pressures.

The effect of repeated assessments on the prevalence of a condition

Higgins' Law states, *"The prevalence of any condition is inversely proportional to the number of experts whose agreement is required to establish its presence."* Let us assume that all observers have to make a certain diagnosis for it to be present. If one observer considers that 100 out of 1000 subjects have angina, a second observer may also agree that 100 have angina but the latter 100 may only include 75 of the initial 100. Twenty-five of both the initial second diagnoses were not in agreement, reducing the confirmed prevalence of angina to 75 in 1000. The use of a third observer may eventually give a prevalence as low as 60 angina cases per 1000. When conducting a long-term clinical trial, it is often essential to have more than one observer but one must realise how the prevalence of a condition, which may be an end point of a trial, may be reduced by repeated examinations.

Good repeatability does not imply validity

The validity of a measurement depends on whether or not it estimates what it is supposed to measure. Random measurements of blood sugar may be examined, blood samples split in two, and the coefficient of variation for a sample computed and found to be satisfactory. However, if random blood sugars are being used to diagnose diabetes mellitus, within-sample repeatability may be satisfactory but between-occasion repeatability may reveal considerable variation. Moreover, a high random blood sugar may not be a good indicator of diabetes mellitus. It may vary according to the recent diet, being high after a carbohydrate meal and low after fasting. A single measurement may have a high within sample repeatability but low validity when used as a diagnostic test for diabetes. A diagnostic test must have both a high validity and a high repeatability and rarely will a test have high validity and low repeatability.

Table 13-1. Some of the detail that may be required in a protocol for a trial where blood pressure is measured.

Machine	Standard sphygmomanometer/random zero, etc
Cuff size	State size of inflatable portion
Cuff-deflation rate	2 mm/second?
Circumstances	Measurement to be made in the doctor's office, laboratory, at home?
Position of patient	Lying, standing, sitting?
Relaxation of patient	Resting for 10, 5, 2 minutes?
Time of day	Morning, afternoon, evening? Certain number of hours after treatment?
Replication of measurements	If replicated, which reading(s) are recorded?
Measurements of diastolic pressure	Point of muffling or disappearance of sound?
Accuracy	Reading to the nearest 2 mmHg?

IMPROVING REPEATABILITY

As discussed in chapter 11, the instructions in the protocol must be clear, the observers fully trained, and the progress of the trial monitored to detect deviations from the trial protocol and other errors.

A clear protocol

Writing a protocol is discussed in chapter 8. The protocol must be detailed and precise, specifying exactly how the trial is to be conducted so that the measurements may be made in a constant manner. For example, if blood pressure is to be measured, the protocol should give the details in table 13-1. For each heading on the left of the table, a statement must appear in the protocol and also in the final report.

Attention to detail is so important that a trial should usually be conducted for a pilot period when the protocol can be amended in the

Table 13-2. Protocol violations observed during a trial involving the measurement of visual acuity before and after refraction, and how the violations were dealt with [273].

Protocol requirement	Violation	Result
1. Visual-acuity lane 20 feet long.	Distance reduced: difficult for clinic to provide distance.	Protocol changed: 10-foot lane.
2. Front illumination.	One clinic provided rear illumination.	Clinic changed practice.
3. Illumination by tungsten spotlights	a. These lamps not used (eg. giving only 10% of specified lighting). b. Bulbs aged and illumination reduced. c. Uneven illumination.	Clinic changed practice. Specially developed visual-acuity boxes issued.
4. Visual acuity defined as that line on which patient gets not more than one letter wrong.	Charts issued had only one 20/200 letter. (Patients could not get this line wrong!)	New charts issued with four 20/200 letters.
5. Refraction.	a. Not done by trained personnel. b. Not done by blind observer. c. Measurement rushed and protocol ignored (or unknown).	Examiners had to be trained and receive certification of training. Technicians trained to take over this measurement from ophthalmologists. Replications done by site visitor.

light of experience (chapter 11). Ferris and Ederer [273] have discussed protocol violations in ophthalmological studies, and table 13-2 reports some of these violations observed during a pilot trial where visual acuity was to be measured. Certain violations led to the clinic changing its practice, some to the protocol being changed and others to the development of special equipment to overcome the difficulties. Many violations would have been expected to reduce repeatability. Of particular interest was the fourth protocol violation, where the patients had to read a line of letters and were judged to be able to do so if they got no more than one letter incorrect. For the line 20/200, the charts included only one letter and in theory, all patients should have been judged 'capable' of reading that line. New charts were constructed with at least four letters per line.

The violation 5.c was also of great interest. A standardised protocol for refracting patients and obtaining the visual acuity had been developed but the techniques were rarely followed. The greatest variability occurred when ophthalmologists were doing the examinations *"A few ... clearly did not know the study protocol at all."* In many clinics technicians were trained to take over the measurements and they adhered strictly to the protocol. The authors suggested that the protocol method was the only way the technicians could make the measurements and therefore they used this method and got reproducible results. The training of observers is of great importance in maintaining the repeatability of the measurements.

Training of observers

Observers must be trained to achieve both repeatable and accurate results. An observer who makes repeatable measurements can still make the measurements consistently too high or too low. An example of the necessity for training is given by the measurement of blood pressure.

Learning the technique
The observer has to be taught which machine to use, how to apply the cuff, as well as the requirements listed in table 13-1. Assuming that the observer is not deaf there are still difficulties in detecting the diastolic sounds. The fourth Korotkoff sound is the point of muffling of sounds (when the sounds stop having a tapping character) and the fifth Korotkoff sound is when the sound disappears. Observers can be trained to recognise these sounds by a sound

film where they listen while observing a mercury
column falling. The observers thus gain
experience and are then tested until they make
the simulated pressure measurements consistently
close to the correct measurement. For real
measurements of pressure a stethoscope can be
adapted so that the trainee and the trainer can
listen to the sounds at the same time and the
training session can continue until consistent
agreement has been reached.

Adherence to the protocol
The observer must be tested for adherence to the
protocol. If the protocol requires blood pressure
measurements to the nearest 2 mmHg then the end
digits 0, 2, 4, 6, and 8 should be recorded with
equal frequency. When monitoring the quality of
blood pressure measurement, a preference for an
end zero is often observed and indicates failure
to adhere to the protocol. However, digit
preference is much less important than an
overreading or underreading due to *observer bias*.
Two machines, used correctly, should prevent this
bias. A random zero sphygmomanometer is available
where pressure is recorded and then the zero
point determined and later subtracted [274].
Digit preference is less obvious with this
machine if zero measurements are made to the
nearest 2 mmHg, but there is nothing to stop
digit preference from the pressures being read to
the nearest 10 mmHg. Moreover allowance must be
made for the fact that the machine consistently
gives lower readings than the standard mercury
sphygmomanometer [275]. The London School of
Hygiene and Tropical Medicine sphygmomanometer
does not allow the observer to watch the mercury
column while listening to the sounds, the descent
of three columns being arrested by the observer
when the changes in the sounds are detected
[276]. The lengths of the mercury columns are
then measured at leisure and observer bias should
be negligible provided that the observer records
the first reading and does not repeat the
measurement. The machine is large and heavy and
has a calibration error that has to be allowed
for [277]. Machines that remove observer bias are
an important advance yet care must be taken to

exclude machine bias. All measurements must be conducted according to a specific protocol and new instruments developed if necessary.

The documents should be designed to prove that the protocol is adhered to; for example, if a tablet count of within ten percent of the expected number is necessary to determine compliance, then the investigator should record the number of tablets rather than whether or not noncompliance was present. Patients often forget to bring their tablets with them but assure the interviewer that they have taken their medication. The investigator will be tempted to record full compliance without the objective evidence required in the protocol. Similarly a protocol may require three blood pressures to be taken and the last to be recorded. In order to check adherence to the protocol, the investigator should be asked to record all three measurements. With increasing adherence the quality of the data will be improved.

Reducing error (quality control)

Should the same or different observers be employed?
Conventional wisdom suggests that if the same observer makes measurements throughout the trial and adheres to the protocol, the measurement technique will be constant and standardised. Alternatively two or more observers can be trained and tested to ensure that they make similar measurements. There are several disadvantages in having only one observer. The observer may be unable to make all the measurements (for example, through illness, holidays, or promotion to another position) and the substitute will be unlikely to have been as well trained as the original observer. In addition, it is difficult for another group to replicate a trial where much depends on the measurement technique of one individual. It would appear safest to train and employ at least two observers.

Errors in recording

Errors in recording will obviously reduce repeatability and the prevention of such errors is discussed in chapter 11. All trial documents must be checked as soon as possible after completion. The design of forms to minimise errors in completion is discussed in chapter 10. At all steps of data collection errors must be avoided. Biochemical and other data must be entered in the units in which they are collected, as it is easier and less prone to error for a computer program to convert data from one set of units to another, than for a busy clinician to do so in his or her outpatient department. The documents must be free of any ambiguity. For example, if the sitting phase IV (muffling of sound) diastolic blood pressure is required, this should be stated in the document as well as in the protocol. Finally, the forms should be designed with efficient data processing in mind but this constraint must not make documents difficult to complete.

The documents for a controlled trial should not accumulate at a local centre until the end of the trial. Either the documents should be checked on the spot or they should be sent directly to a coordinating centre. Whatever the method of monitoring, *quality control* must be adequate.

Errors in diagnosis

It cannot be assumed that all diagnoses made during the trial are correct. Kahn and his colleagues [278] provided some very interesting examples of diagnostic difficulty in glaucoma and concluded that more training, more tightly written protocols, and an increase in the use of objective measurements are all required. It is also important that the criteria on which a diagnosis is based are recorded and not simply the diagnosis.

Errors in measurement

Errors of measurement may be random or systematic. Both are important, and the latter (systematic error or bias) can produce a false positive result if it occurs in one treatment group more than another. Before commencing a trial, the investigator will have some knowledge of the variability of the measurements to be made. This information may be necessary for the calculation of the numbers required for a particular trial and will be required for quality control during the trial. The acceptable range of results for every measure must be defined and the documents examined to detect results that lie outside these limits. The outlying measurements must be investigated further.

Clerical errors
Clerical errors are mainly those of transferring information and four have been defined [279]: person to document; instrument to document; document to computer and computer output to report. We can add another transfer error, typing mistakes in any of the draft or final reports.

The solution to these errors is repeated checking. Ideally two persons should be responsible for the completion of all documents and data entry must be verified by re-entry. In the latter process the data are rejected when the data from the first entry do not coincide with the second keyboard input. Despite all care, errors still occur and all data have to receive two examinations: first are the data within the range of possible answers and second do the data have internal consistency? Examples of consistency checks are given by the following: a male patient who is pregnant; a 50 kg person losing 20 kg in weight; a woman complaining of impotence; and a patient's ageing by more than one year over a one-year period.

The reduction in variability achieved in a cross-over trial

In a cross-over trial a patient takes one treatment, then a further treatment or treatments. The patient acts as his own control and the variability of the response to the treatment is correspondingly reduced. In certain instances the reduction in variability may allow a result to be achieved with a fraction of the number of patients. The advantages and disadvantages of cross-over trials are discussed in chapter 5.

VALIDITY OF MEASUREMENTS AND TRIAL RESULTS

The validity of a measurement is the extent to which it measures what it is supposed to measure. This can be extended to consider the validity of a trial result. One dictionary definition of validity is *"so executed etc, as to have binding force"* [280]. A trial result may not have binding force when executed incorrectly or when the trial result is not interpreted correctly. An incorrect interpretation can arise when an investigator decides that the same result will be true in different subjects (extrapolation of trial results) or when the trial reveals a particular result and the investigator jumps to a different conclusion.

	Patients' answers	
	Yes (have been admitted with heart attack)	No (have not been admitted with heart attack)
Known to have been admitted with a heart attack.	a	b
Known to have been admitted with a different diagnosis.	c	d

Figure 13-4. The validity of self-administered questionnaire to determine whether or not a patient's recent hospital admission had been due to a heart attack. The letters a, b, c, and d are numbers of patients.

Validity of the measurement

Figure 13-4 provides a theoretical example in which the validity of a question, *"Have you been admitted to hospital with a heart attack during the last six months?"* is examined. The question was asked of a group who had been admitted with myocardial infarction and a group that had been admitted with a variety of other medical conditions. Their answers are entered in the columns of the figure and cross-tabulated according to the known truth. If the total number of patients was n then the proportion of correct responses was $(a + d)/n$. Similarly the false negative rate can be calculated as $b/(a + b)$ and a false positive rate as $c/(c + d)$. Such a tabulation could be used to examine the validity of other data. For example, the columns could be a normal or high random blood sugar and the rows the presence or absence of diabetes mellitus as determined by a glucose tolerance test. Similarly, electrocardiographic abnormalities, or serum enzyme changes, may be examined separately as tests for acute myocardial infarction provided a definitive diagnosis or gold standard can be provided from, say, coronary angiography.

Validity of trial results

Subjects of a different age, sex, or race
The response to a treatment may vary between men and women, alter with age, and differ between the races. We cannot assume that the results of a trial including, for example, only young men are applicable to elderly women. This mistake commonly arises when trials are conducted on laboratory staff (these persons tending to be young and male), and an example of this problem is often provided when the clinical pharmacology of a new drug is determined on volunteers.

Even after allowing for age, sex, and race the selection of subjects into a trial makes it difficult to apply the results to a superficially similar group. With patients, selection into the trial may favour those with either a particularly severe or mild form of the disease. Also, patients taking part in clinical trials are often more compliant with therapeutic advice than patients subsequently offered the treatment. Lastly, a trial result may occasionally depend heavily on a subgroup of subjects.

The severity of the disease
The Veterans Administration Co-operative Study Group on Antihypertensive Agents trial [64] admitted only men who were patients attending Veterans Administration hospitals; these men had moderate to severe hypertension. First, the level of diastolic blood pressure leading to entry to the trial was from 90-129 mmHg after *four to six days resting in hospital*. This level of hypertension had also to be observed in an outpatient department after a two-to-four-month period of placebo medication, at which time the pressure was designated the untreated pressure. In addition the diastolic blood pressures were measured at the lower point of disappearance of the Korotkoff sounds, not the point of muffling of these sounds.

A patient in a doctor's office with a similar level of untreated pressure may not be comparable to those subjects in the Veterans Administration trial. He may prove not to have hypertension after hospital admission or blood pressure may fall during prolonged placebo therapy. Also, the point of muffling of sound may be used to determine a (higher) diastolic pressure in the doctor's office. A patient with a casual diastolic pressure above 100 mmHg may therefore be equivalent to a patient with a diastolic blood pressure of only 90 mmHg in the Veterans Administration trial. If the doctor remembers to allow for the possible effects of hospitalisation, placebo, and different measurement techniques he must also recall that 58 percent of the Veterans Administration patients had been categorised as having a preceding cardiac, central nervous system, or renal abnormality [281]. The doctor must ask himself how valid are the results of the trial for the patient confronting him in his office.

Table 13-3. Criteria for withdrawal from the Veterans Administration Cooperative Study Group on Antihypertensive Agents trial during the placebo run-in period.

1.	Failure to appear for a regularly scheduled clinic appointment
2.	Failure of the urine to contain the prescribed placebo
3.	Failure on a tablet count
	a. Over 10% too many tablets left
	b. Five percent or more too few tablets left

Subject cooperation and compliance

Those entering a trial tend to be cooperative, willing to make several visits, undergo investigations, and sign consent forms. They are probably more likely to adhere to therapeutic advice than the average patient. The adherence to advice has been termed *compliance* (chapter 11) and a trial result may not be valid for a more representative group of patients including a high proportion of noncompliant individuals. In addition, a trial may be specifically designed to exclude noncompliant patients. Table 13-3 gives the withdrawal criteria for the Veterans Administration trial discussed above. Nearly half of the eligible patients were excluded for not attending a clinic appointment, or for not having any placebo marker in their urine, or for failing to produce an approximately correct number of remaining placebo tablets. The trial was a test of active treatment in those who were prepared to take antihypertensive medication regularly and the results are valid for those who do so. This is not a criticism of the trial but there is a limit to how far we can generalise from a particular group of patients to the general population. The results for male veterans are not necessarily valid for women, nor are the results applicable to patients who usually forget to take their tablets. From the results of such a trial, it will be difficult to estimate the response in a group of patients including noncompliant persons. This problem is discussed more fully in chapter 11.

*Results are only applicable to certain subgroups of
patients in the trial*
A difficult problem is provided by a trial with
a significant overall result that relies heavily
on some subgroups and not others. For example, in
the Hypertension Detection and Follow-up Program
trial [282], an overall beneficial result was
observed but not for the subgroup of white women.
It may be dangerous to generalise from the
overall result to white women when such subgroup
differences are apparent.

The investigator over-interprets the results
We shall consider a hypothetical trial that
provides an example of a non sequitur. The trial
is designed to determine whether or not a
particular pharmaceutical agent can stop
cigarette smokers from indulging in their habit.
The trial shows that the drug stops smoking in a
significant proportion of smokers. However, the
authors know that smoking causes heart disease
and conclude that the drug will reduce total
cardiovascular mortality. This does not follow
because when a causative factor is removed the
previous adverse effects may persist. A different
trial is required to show that stopping smoking
prevents cardiovascular death and, in fact, one
trial of antismoking advice demonstrated that
stopping smoking was associated with a reduction
in respiratory symptoms but could not prove a
reduction in mortality over the short period of
the trial [164].
    Similar examples in the cardiovascular field
are provided by trials showing that
antihypertensive drugs lower blood pressure. Such
trials do *not* prove that the drugs reduce stroke
mortality. The results of a trial must not be
extrapolated beyond the observations. It is not
valid to conclude that a treatment has wider
effects than those observed.

CONCLUSIONS

This chapter has described how the precision of
repeated measurements can be estimated together
with their accuracy in comparison with a known

standard. The relationship between bias, repeatability, and validity has also been discussed. The concept of quality control was introduced and the methods to achieve fewer errors considered, a clear protocol and careful training of the observers being the most important. The detection of errors was reviewed.

The results of a trial may be invalid if the trial is not performed correctly. The provision of adequate controls, avoidance of bias in the results, and reduction in variability are discussed in chapter 7. In addition, the results of the trial must not be extrapolated beyond the observations made and some examples have been presented in this chapter. The investigator must refrain from jumping to conclusions that he cannot support and must not assume that the results of his trial are valid for subjects of a different age, sex, race, disease severity or for noncompliant subjects. However, the results of a particular trial can be expected to be repeatable for a demographically similar group of subjects selected and treated in an identical fashion. The results should be valid for such a group.

APPENDIX

Calculation of the precision and accuracy of measurements

Measurements of systolic blood pressure taken with a mercury sphygmomanometer were compared with two prototype instruments. The first Latin square illustrated in figure 5-3 (chapter 5) was utilised to determine the order. The standard machine was A and the prototypes B and C. Three patients were randomised as patients 1, 2 and 3.

The results for three patients cannot be conclusive but the prototype machines tended to have a worse (higher) precision or standard deviation values than the standard machine. Moreover, the accuracy of the new machines was not acceptable.

		Machine		
		A	B	C
Patient 1		168	183	164
		153	168	168
		151	174	165
	Mean	157.3	175	165.7
	SD	9.3[†]	7.5	2.1
Patient 2		149	172	169
		149	164	178
		150	164	174
	Mean	149.3	166.7	173.7
	SD	0.6	4.6	4.5
Patient 3		123	110	088
		118	109	110
		118	125	110
	Mean	119.7	114.7	102.7
	SD	2.9	9.0	12.7
Overall	Mean	142.1	152.1	147.4
	Precision	5.6[‡]	7.3	7.9

[†] Standard Deviation (SD) =

$$\sqrt{\frac{(168 - 157.3)^2 + (153 - 157.3)^2 + (151 - 157.3)^2}{3 - 1}}$$

$$= 9.3$$

[‡] Precision

$$= \sqrt{\frac{9.3^2 + 0.6^2 + 2.9^2}{3}}$$

$$= 5.6$$

Accuracy of machine B = 175 - 157.3 = +17.7 mmHg for patient 1. +17.4 mmHg (patient 2) and -5.0 mmHg for patient 3.

Overall accuracy of machine B (irrespective of sign)

$$= \frac{17.7 + 17.4 + 5.0}{3}$$

$$= 13.4 \text{ mmHg}$$

14. ANALYSIS OF THE RESULTS, SUBGROUP AND META-ANALYSIS

Several authors have reviewed the statistical analyses appearing in reputable medical journals. The reports reveal a record of error and incompetence [283-289]. This chapter considers these problems and the statistical principles required in the analysis and interpretation of the results. It is beyond scope of this book to describe in detail the statistical methods required to analyse the results of randomised controlled trials. The reader is referred to standard texts such as the books by Armitage and Berry [290], Everitt [291] and Bailar and Mosteller [292].

Two strategies have been suggested to improve statistical analyses: first, a statement should be made in the protocol of the analyses that are intended. This can be scrutinised by fund-giving agencies and ethical committees [289]. The investigator writing the protocol must have a grasp of statistics or seek the help of a statistician. The second strategy is to have all articles scrutinised by a statistician prior to acceptance for publication. In the past it appears that papers with a detailed description of statistical methods have been referred for a statistical opinion whereas those with little or no mention of these methods have not. The latter articles have proved to be those that actually needed the statistical review [284]. It must be noted that it may be too late to seek the help of a statistician at the report writing stage. Statistical expertise should be available at the design stage.

When writing the protocol and requesting funding for a trial, care must be taken to ensure that adequate arrangements are made for prompt data processing and analysis as it may be unethical to proceed with a trial longer than necessary. Peto and his colleagues also considered prolonged trials and stated, *"Collect as much data as possible at first presentation, only data which are strictly necessary thereafter, and analyse the data you do collect very thoroughly"* [293]. The initial data determine whether randomisation has been successful in producing equivalent groups and may also be of use in determining prognostic factors.

The present chapter considers errors in analyses; checks on randomisation; analysis of normally distributed data, proportional data, and survival data; confidence limits; and the problems posed by defaulters. Subgroup analyses and meta-analysis are also briefly discussed.

COMMON ERRORS DURING ANALYSES

The errors frequently encountered in reports of clinical trials include: confusion about the experimental unit in the trial; the failure to use a statistical test; failure to state the statistical test employed when one is used; the inappropriate use of t tests; the reporting of standard deviation instead of standard error of the mean and vice versa; the inappropriate use of one-sided significance tests for α; confusion over the meaning of P; and failure to analyse defaulters in a reasonable manner.

Confusion about the experimental unit

It is not always individuals that get randomised in clinical trials; a bodily part may be randomised, for example an eye, limb, or joint. In patients with rheumatoid arthritis it is possible to randomise joints to one of two treatments and five patients may have, say, 16 joints randomised for treatment. The most extreme example of more randomisation than subjects is the single-person trial (chapter 5). In this type of trial episodes of illness are randomised for treatment (for example, recurrent episodes of hay fever or asthma). In one trial a patient with suspected myasthenia gravis received in random order, placebo, prostigmine, and D-amphetamine [294]. This trial was of diagnostic value in this patient but would a trial of treatment in one person be likely to have any general applicability? For example, a trial of hay fever treatment in one patient demonstrates that drug A is statistically significantly more effective than drug B. What can we deduce from this trial? We can only conclude that drug A is to be preferred in this one subject. We cannot conclude that, in general, patients should be given drug A and we are mainly concerned with the overall validity of our results (chapter 13).

When more than one patient is involved in a trial but there are more randomisations than patients we may get the results given in table 14-1. In this hypothetical trial 11 children had a total of 66 warts but one child had 22 warts and the others had seven or less [295]. Taking a wart as an experimental unit, treatment A cured significantly more warts (62 percent) than B (28 percent), $P < 0.01$ using a chi-square test. However, the data show that this result was due to the excellent result in the child with 22 warts. If the experimental unit is taken to be a patient, then four children responded more to A than B, two more to B than A, and five responded

Table 14-1. The results of a trial of the treatment of viral warts in 11 children. The apparently beneficial results of treatment A were almost entirely due to the effect of this therapy in child number 6.

Child	Number of warts randomised to treatment A	Number of warts cured	Number of warts randomised to treatment B	Number of warts cured
1	3	0	1	1
2	1	1	1	1
3	3	3	4	0
4	2	1	1	0
5	2	2	2	2
6	11	10	11	0
7	4	0	3	3
8	1	0	2	0
9	1	0	1	0
10	3	2	3	0
11	3	2	3	2
Total	34	21 (62%)	32	9 (28%)

similarly - hardly an impressive result. In clinical trials the experimental unit must be the subject and the results arranged so that each subject is equally important. In any statistical test the degrees of freedom should be less and never more than the number of subjects.

A statistical test is not employed

Contemporary reports of clinical trials usually employ a statistical test as most authors are convinced of the necessity of supporting their conclusions with tests of statistical significance. However, if the result of treatment is very remarkable, such as the survival of the first patients with tuberculous meningitis to be given streptomycin, then neither a controlled trial nor statistical analysis may be required. Unfortunately most treatments are not so effective, the outcome of the untreated condition is more variable, and both controlled trials and statistical evaluation are required.

The statistical test is not identified

Three reviews have reported that about a third of publications state a P value without quoting the procedure used to test for statistical significance [284,289,296]. It is of great importance to state the method used to arrive at a conclusion in order that the readers may fully evaluate the results and assess the suitability and accuracy of the statistical tests.

The statistical test is used inappropriately

Glantz [289] has demonstrated that at least a third of articles quoting the results of more than one t test should have employed a test that allowed for the use of multiple comparisons.
Let us consider the results given in figure 14-1a. Four drugs are compared: A, B, C, and D. Six t tests are reported in order to test for differences between the four mean results. By employing the six tests (t_1 to t_6) all results are compared and yet we know that extreme results must not be selected and contrasted. When only

a)

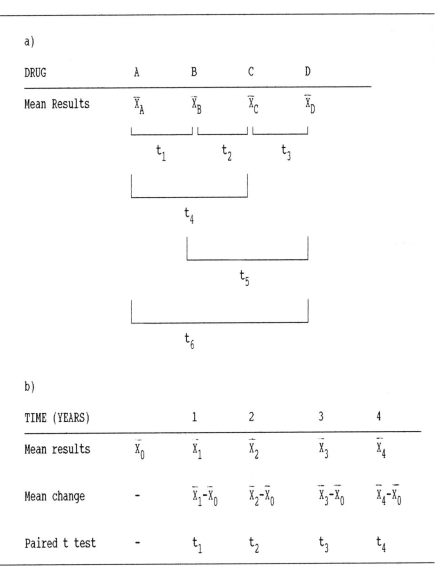

DRUG	A	B	C	D

Mean Results $\bar{X}_A$ $\bar{X}_B$ $\bar{X}_C$ $\bar{X}_D$

t_1 t_2 t_3

t_4

t_5

t_6

b)

TIME (YEARS)		1	2	3	4
Mean results	$\bar{X}_0$	$\bar{X}_1$	$\bar{X}_2$	$\bar{X}_3$	$\bar{X}_4$
Mean change	-	$\bar{X}_1-\bar{X}_0$	$\bar{X}_2-\bar{X}_0$	$\bar{X}_3-\bar{X}_0$	$\bar{X}_4-\bar{X}_0$
Paired t test	-	t_1	t_2	t_3	t_4

Figure 14-1. Results in two trials where multiple t tests should not be employed.

Figure 14-1a. Gives the mean results (X) for four treatments A, B, C, and D. t_1 to t_6 indicates the multiple t tests that could be (but should not be) employed.

Figure 14-1b. Provides the mean results for one group of a within-patient trial given a single treatment for four years. $\bar{X}_0$ is the mean baseline result. t_1 to t_4 are the four paired t tests that could be performed on the change in the variable. Such multiple testing should be avoided (see text).

one *t* test has to be performed in a test for a five percent level of significance, we may be willing to accept a one in 20 chance that we have incorrectly rejected the null hypothesis. If we perform six *t*-tests we approach a one in four probability of a wrong conclusion. The probability is not as high as 6 X 5% = 30%, but is given by the formula:

Probability of making a mistake = 1 - probability of not making a mistake .... (14.1)

$= 1 - 0.95^6$

$= 1 - 0.74$

$= 0.26$

or a 26% chance of making an error [297].

If there were ten drug groups it would be more obvious that we cannot select the smallest and the largest result and compare the two using a straightforward *t* test. The solution to the problem is to select a test that takes into account the multiple comparisons performed: for example, a studentized range test or Scheffe's test [290], or, where appropriate, to perform an analysis of variance.

Figure 14-1b illustrates the situation where one group in a trial is given a particular treatment and the average result of a variable, say, weight, is $\bar{x}_0$ at baseline and $\bar{x}_1$ to $\bar{x}_4$ after 1, 2, 3, and 4 years, respectively. The investigator wishes to see if there has been any statistically significant change in weight after any one of the intervals of time and the change is best examined using a paired *t* test. Unfortunately this gives rise to four tests of significance, and we have the problem of multiple testing and selecting the extremes. Analysis of variance would be an appropriate solution in order to examine the data for a significant change after any one of the four years. The analysis can even be extended to detect a tendency for weight to alter progressively and linearly with time.

Multiple testing will also be a problem if we examine changes in, say, 20 variables, but confine the results to those at four years. Examining many variables increases the probability of a significant result. If each variable is independent of the others we expect, at the 5% level of significance, 20 x 0.05 = 1 result significant at the 5% level [298]. We must therefore demand a high level of significance and the Bonferroni method suggests dividing the α level by the number of tests to decide on the new α level eg 0.05/20 = 0.0025 [299]. It must be stressed that this approach is not applicable if the variables are related and not independent. For example, if we test changes in both diastolic and systolic blood pressure, we would expect both to change in a similar direction and dividing by the number of tests would not be appropriate. One possibility in this situation is to demand a higher than normal P value, say $P < 0.01$.

One-tailed tests of significance

In biological work, the test for a type I (α) error must be two-tailed (chapter 6) as the result may either be in the expected or the unexpected direction. The only exception to this rule is the decision-making trial (for example, a trial to determine whether or not a pharmaceutical company should investigate a new drug). In this instance the firm is not initially concerned with whether or not the new treatment is the same as a control treatment, but only whether the company should investigate the new treatment or not. However, in the usual report of an explanatory trial, if the reader detects a P value based on a one-tailed test this should be multiplied by a factor of two to provide a two-tailed assessment.

The meaning of P

When a statistical test is performed on a difference between two groups, P is the probability that the particular difference would be explained by sampling error and observed if the null hypothesis is true and there is actually no difference. *P* is *not* the confidence with which we accept that the difference is exactly zero. For example, if *P* = 0.45, we cannot reject the null hypothesis. Forty-five repeat samples out of 100 would be as extreme as this or greater and 45 percent is the probability of making a type I error and rejecting the null hypothesis when we should not. However, *P* is not the probability of making any kind of error; we can also make an additional error, type II (ß) and support the null hypothesis when there actually is a true difference between the treatments (chapter 6). A large *P* value does not imply that a true difference has been excluded. Freiman and colleagues [288] examined 71 negative randomised controlled trials mainly published in the *Lancet, New England Journal of Medicine,* and the *Journal of the American Medical Association.* Half the trials had more than a 74 percent probability of failing to detect a 25 percent improvement with therapy. When reporting the results of controlled trials the confidence limits should be reported for any result (see below).

Standard deviation (standard error) of the mean

The calculation of a standard deviation was described in chapter 6. The standard deviation of a series of normally distributed results summarises the spread or variability of the data. As Glantz stated, *"When observations are equally likely to be above or below the mean and more likely to be near the mean than far away, about 95% of them will be within 2 standard deviations on either side of the mean"* [289]. A standard deviation therefore summarises the data and is an accepted description of the spread or variability of the raw data. The variability of a mean result is given by the standard deviation of the mean which is known as the *standard error of the mean* (SEM) and is calculated by:

$$\text{SEM} = \frac{\text{Standard deviation of the raw data}}{\sqrt{\text{number observations}}} \quad ..(14.2)$$

Glantz concluded, "... *the standard deviation, not the standard error of the mean, should be used to summarise data.*" This is true when describing the information on subjects before entry to the trial but not necessarily for the results of the trial. The outcome of a trial will usually consist of the difference between the mean results of the separate treatment groups. The variability of this difference will depend on the standard error of the means and not simply on the standard deviations of the raw data.

Assume drug A produces a mean result of $\bar{X}_A$, standard deviation, SD_A and standard error of the mean, SEM_A. Similarly drug B produces mean $\bar{X}_B$, SD_B, and SEM_B and a comparison of the two means is made by a t test calculated as follows:

$$t = \frac{\bar{X}_A - \bar{X}_B}{\text{standard error difference}} \qquad \cdots\cdots\cdots(14.3)$$

$$t = \frac{\bar{X}_A - \bar{X}_B}{\sqrt{SEM_A^2 + SEM_B^2}} \qquad \cdots\cdots\cdots\cdots\cdots\cdots(14.4)$$

The value of t could be more accurately calculated by pooling the separate variances [290]. A report of a trial should therefore include the appropriate means and standard errors of the means (or standard deviations and the numbers of measurements involved) so these calculations can be made.

In a graphical representation of the outcome, a mean result is often represented as ϕ, and the legend to the figure must always state what the bars represent. Traditionally the figure indicates the mean and *one* standard error of the mean on either side. It should be noted that this figure does not provide the confidence limits for the mean and it is surprising that such a device has achieved widespread acceptance.

The handling of data on dropouts and withdrawals from the trial

The inclusion or exclusion of withdrawals from the trial often leads to errors in analysis and is discussed later in this chapter.

ANALYSES TO DETERMINE THAT RANDOMISATION HAS PRODUCED EQUIVALENT GROUPS

The characteristics of the patients at presentation should be compared between the treatment groups in order to evaluate the results of randomisation. For each normally distributed characteristic the mean, number of measurements, standard deviation (rather than SEM as the raw data are being described), and range should be presented. For discrete data the proportions must be reported (for example, the percentage male, married, or black). Usually statistical tests are conducted to see if there are differences between the groups at the start of the study. However, Altman has pointed out [300] that performing a significance test to assess the *"probability of something having occurred by chance when we know that it did occur by chance... is clearly absurd"*. However he does agree that we must establish whether or not randomisation has been satisfactory and statistical testing is one method. In blinded randomised trials Chalmers *et al* [301] found 3.5% had baseline variables different at the 5% level of significance against 7% for randomised but not blinded trials. The expected number should be 5% suggesting that trials with unblind randomisation may be open to some bias.

Altman suggests making an initial comparison *"non statistically using a combination of clinical knowledge and common sense"* [300]. If a variable differs between the groups and may be related to the outcome (a confounding variable), the analysis should adjust for this possible imbalance. This strategy of ignoring *'P'* values has the benefit of assuring that an adjusted analysis is performed when the comparison is not statistically significant. This is very necessary when the difference is clinically important. Lavori *et al* [302] describe three methods of adjusting the outcome measures: stratifying by the confounding variable (most useful for a categorial variable such as gender); direct standardisation; and modelling. It must be remembered that these adjustments increase the precision of the treatment effects and may increase or decrease the apparent effect of treatment. A delicate balance has to be drawn between the reader rejecting the analysis because of imbalance in baseline variables and the reader rejecting the results because of the complexity of the analysis. A simple analysis should be provided first and the adjusted treatment effects provided as additional information.

The following concluding remarks should be noted:

1. When a large number of characteristics are compared, one may well differ between the groups by chance alone.
2. When large numbers of patients are entered into a trial, some differences between the groups may be statistically significant but they may not be large or of biological importance.
3. When the trial includes a small number of patients, significant differences are likely to be of biological importance and even nonsignificant differences may be large.
4. If more statistically significant differences between the groups are detected than would be expected by chance, the randomisation process may have failed and a biased selection into the different groups may have occurred.
5. Any random differences between the groups can be adjusted for retrospectively in the analysis [293].

ANALYSIS OF THE RESULTS: QUANTITATIVE OR CONTINUOUSLY DISTRIBUTED DATA

The data must be checked for outliers and distribution, transformed if necessary, and subjected to the appropriate statistical test.

Data checking

It is of the utmost importance to rectify any data that have been incorrectly recorded and to eliminate any results that constitute errors in measurement. However, great care must be taken that outliers are not removed in order to improve the results and support the investigators' preconceived notions.

The data are checked by examining the frequency distribution for impossible or outlying values. These are known as range checks. In certain situations, consistency checks can be performed, for example, to confirm that all patients who are pregnant are female, and that systolic pressures exceed diastolic on a plot.

When computing, statistical packages such as SPSS (Statistical Package for the Social Sciences) [303] can be used to derive a frequency distribution, mean, and standard deviation. Healy and others have described rules for the detection of outliers [304-305].

Tests to confirm that the data are normally distributed

Normally distributed variables are continuous quantitative measurements (for example, blood sugar) that conform to various tests of distribution. The most important test is the degree of skewness that can be assessed statistically and also examined by plotting the frequency distribution of the variable. An example of a skewed distribution is provided by the frequency of plasma urea. This has a right-hand tail of high values. Plasma urea has to be transformed to achieve a normal distribution, but the distribution of blood sugar is not markedly skewed.

The normal distribution is also bell-shaped and a test for kurtosis indicates whether the data have too flat (uniform) or too peaked a distribution.

Transformation of the data

The most efficient ways of comparing continuous data are the *t* test and the analysis of variance. These analyses require a normal distribution but data that are positively skewed (to the right), such as plasma urea, can be rendered more normal by a logarithmic transformation. The distribution of the transformed variable should be examined to confirm that the skewness has been reduced.

The data cannot be transformed to a normal distribution

If the data do not conform to a normal distribution, they can be analysed using nonparametric statistical techniques, often with little loss of efficiency [290-292]. However, statistical techniques based on the normal distribution are most efficient and are to be recommended when possible. When the change in a particular variable is to be analysed, the distribution of the original data may not be important if the change in the variable is normally distributed.

The tests to be used with normally distributed data

The *t* test and analysis of variance should be employed where appropriate. The Student's *t* test is a well-known statistical test and the analysis of variance (ANOVA) is becoming more familiar and better understood by the average reader of trial results. ANOVA may prove that the result of one or more treatment groups is significantly different from the other groups. The problem will remain of determining which groups differ from the others. The Student's range and other tests can be used to determine exactly which between-group comparisons are not compatible with the null hypothesis [290].

a)

	TREATMENT A	TREATMENT B	TOTAL
NO CHANGE	a	b	a+b
CURED	c	d	c+d
TOTAL	a+c	b+d	n

$$\chi^2 = \frac{[(ad-bc) - \tfrac{1}{2}n]^2 n}{(a+b)\,(c+d)\,(a+c)\,(b+d)}$$

b)   AFTER 1 YEAR

		CONDITION +	CONDITION -	
T				
O				
E	CONDITION +	K	r	
N T				
T R	CONDITION -	S	m	
R I				
Y A				
L				T

<u>McNemar's test</u>

$$U = \frac{r - \tfrac{1}{2}n}{\tfrac{1}{2}\sqrt{n}} \qquad \text{where } n = r+s$$

Figure 14-2. The analysis of proportional data. **Figure 14-2a.** Two different groups of patients receive different treatments. The chi=squared statistic assesses the results of the trial and the answer should be compared to tables of X^2 on 1 degree of freedom. n = a+b+c+d.

Figure 14-2b. A group of T patients is examined to see if there is a change in a condition from entry to trial to a one-year assessment. McNemar's test provides the approximate number of standardised normal deviates (U) that can be examined in the appropriate table.

ANALYSIS OF THE RESULTS: QUALITATIVE OR DISCRETE DATA

Proportional data, such as the percentage cured, improved, or dead, require statistical tests such as the chi-squared (X^2) test. Figure 14-2a gives the calculation of the usual chi-squared test which incorporates a continuity connection. Tables of chi-squared must be consulted to determine whether or not the result exceeds a given level of significance. When examining the change in a proportion over a period of time in the same subjects we have to perform an analysis suitable for paired data. Figure 14-2b gives the calculation for McNemar's test [290], and the result of the test has to be examined in tables of standardised normal deviates. McNemar's test is useful in determining whether any significant change is occurring within one treatment group rather than comparing between groups.

ANALYSIS OF THE RESULTS: SURVIVAL DATA

The survival from a rapidly fatal disease in two groups can be compared using a chi-square test as above. However, when the disease is not quickly fatal we will wish to take into account, not only the fact of death, but also the length of time before death. Also, certain patients may be lost to follow-up, and some allowance has to be made for this fact in the calculations. The appropriate analysis is the construction of a life table.

The life table

A life table is best represented as a graph in which the proportion of survivors over time is plotted. The technique for constructing a life table is well described in Armitage and Berry's book on medical statistics [290] and in a useful article by Peto and his colleagues [293]. The data required on each patient for such an analysis include the date of randomisation; the date of completion of the study; whether the patient is dead or alive; if dead, the date of

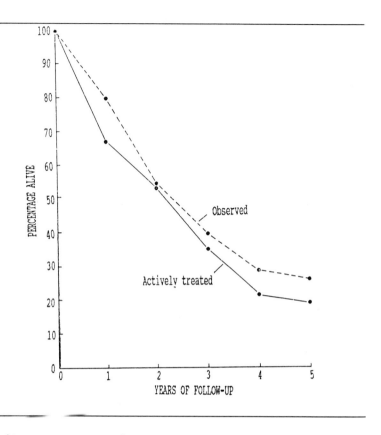

Figure 14-3. Life-table representation for 60 very elderly hypertensive subjects treated with methyldopa and 60 persons who were simply observed [306].

death; and if the patient is lost to follow-up, the date the patient was last known to be alive.

Statistical tests of survival

Statistics such as the median and average survival times can be very inaccurate unless nearly everyone dies and the data are complete. Treatments may also differ in their acute and long-term effects; for example, the life table may show a benefit for one treatment after one year but this may be reversed after, say, three years. A statistical test is required to compare the whole of the two survival curves illustrated in figure 14-3. Peto and colleagues have come down firmly in favour of the log-rank test [307,308] and have described how to perform the

this test [293]. The results given in the figure
are straightforward in that the difference
between the two groups is not significant after
five years (0.3 < P < 0.5). The confidence limits
are discussed below.

Life-table analysis to determine prognostic features

The life table may also be useful when
subdividing the results according to suspected
prognostic features but within particular
treatment groups. For example, patients given
placebo may have two survival curves constructed,
one for smokers and one for nonsmokers. The
comparison of the curves will indicate the
prognostic effect of smoking. More importantly
the life-table analysis may be adjusted for
baseline variables, an important procedure if
there are inequalities following randomisation.

Cox [309] developed a multiple regression
approach to use with censored data, the
Proportional Hazards model. Although other
methods have been subsequently proposed, Cox's
model remains the most widely employed and most
informative method [310]. The major statistical
packages include programs for the necessary
computations. The information required includes
the start date for the individual subject, the
date of an end-point (if any), their date of
censoring if they are censored (otherwise the
date of the end of the study), and the variables
to be considered as risk indicators.

Survival curves employing the Proportional
Hazards model may be compared for statistical
significance [311].

THE CONFIDENCE INTERVAL MUST BE GIVEN FOR ALL TRIAL RESULTS

The researcher may summarise the result of a
trial as *"the difference between the treatments was not
significant, P >0.05"* or *"the difference was significant,
P <0.05"*. This is not satisfactory and the precise
P value should be quoted to at least 2 decimal

places, but even then it gives no information on the actual differences observed or on the range of possible differences between the groups. The confidence interval gives this range.

Medical research is concerned as much with estimation as with hypothesis testing and, since repeated studies would provide different estimates, we need to know an interval of values that is likely to contain the true result [312]. Our sample statistic - for example, the mean difference - is our best estimate of the result. However, this figure is only a single result. If a second, third, or even a hundred such comparisons were made, what would be the results? The confidence limits (CL) for one comparison encompass a range of values (the confidence interval [CI]) that is likely to cover the true population value.

Definition of a confidence interval [313]

The confidence interval is a range of values that is likely to cover the true but unknown value. In classic significance testing the confidence interval is based on the concept of repeated trials or studies - ie, with the 95% confidence interval, if the study was repeated 100 times, the confidence interval would be expected to include the true value on 95 occasions. However, a reported CI either does nor does not include the true value.

Use of the confidence interval

The confidence interval is important when a statistically significant result is reported and essential when a negative result is described. If a trial demonstrates a positive result - say, a rise in haemoglobin averaging 1gm/100ml and the result is statistically significant at the 5% level - then we know that within the 95 percent bounds of probability the result was not compatible with a zero increase in haemoglobin. However, the range of increase compatible with a probability of 0.95 may be large: for example, +0.2 to +1.8gm/100ml. The limits of this range are the 95 percent confidence limits. However, the term confidence interval rather than confidence limits is to be preferred. The term 'limits' implies that values are not possible beyond these and ignores the fact that in 5 out of 100 studies the true values would be outside these limits.

The importance of reporting the confidence interval was stressed by Wulff in 1973 [314] and supported since then by many authors [125,313, 315]. Gardner and Altman [316] have formulated a policy for the *British Medical Journal* – namely, that *"Confidence intervals, if appropriate to the type of study, should be used for major findings in both the main text of a paper and its abstract"*.

The confidence interval for normally distributed data

The confidence limits for normally distributed data are simple to calculate because standard errors are calculated as part of the test of statistical significance. The 95 percent confidence limits for the mean difference (MD) between two treatments in a large number of patients are given by the following formula:

95% CL= MD - (1.96xSED), MD + (1.96xSED)...(14.5)

where SED is the standard error of the difference between the two means. The difference between the limits gives the 95% confidence interval (CI). For samples less than 30, the multiplier is greater than 2 and must be taken from tables of the t distribution with n - 1 degrees of freedom, where n is the sample size [290].

The SED differs according to whether the differences are paired, as in a within-patient or paired analysis, or whether the average difference is simply the difference between two means (an unpaired analysis). Gardner and Altman [316] have clearly shown how to make these calculations and have provided examples.

Confidence intervals for proportional data

For data in a proportional form the standard error of the difference has also to be calculated. Baber and Lewis [317] reported the confidence limits on a series of trials of the secondary prevention of myocardial infarction. They calculated the SED and 90% confidence limits as follows:

let n_1 and n_2 be the number of subjects in two treatment groups.
let p be the overall death rate and p_1 and p_2 be the death rates in the two treatment groups.
let q be the overall survival rate = $1 - p$

then

$$p = \frac{n_1 p_1 + n_2 p_2}{n_1 + n_2}$$

$$\text{SED} = \sqrt{pq \left(\frac{1}{n_1} + \frac{1}{n_2} \right)} \quad \dots \dots \dots \dots \dots \dots \quad (14.6)$$

and

$$90\% \text{ CL} = (p_1 - p_2) \pm \left[(1.64 \times \text{SED}) + \tfrac{1}{2} \left(\frac{1}{n_1} + \frac{1}{n_2} \right) \right]$$

where $\tfrac{1}{2}(1/n_1 + 1/n_2)$ is a correction factor as the data are not continuous.

Again Gardner and Altman have provided a worked example of how to derive the CI in the paired and unpaired case [316].

Confidence intervals for other outcome measures

Methods have been reported that provide CIs in regression analysis [290], in determining risk ratios and relative risks [318], when non parametric methods are employed [319] and when life tables are employed [308,309,320]. After five years the 95 percent confidence limits for the negative results reported in figure 14-3 ranged from a benefit due to treatment of 12 percent and an increased mortality of 36 percent, indicating that the trial was far too small.

Confidence limits when the result of the trial is negative

When a trial reports a near zero effect the confidence limits will indicate the extend to which this result is compatible with a benefit from a treatment in one direction and an adverse effect in the other direction. Of 16 trials with negative results analysed by Baber and Lewis [317] 12 (75 percent) results were compatible with a treatment effect reducing mortality by 50 percent and eight (50 percent) were compatible with an increase in mortality of 50 percent. Gore has also provided very useful examples of the place of confidence limits in assessing the results of clinical trials, including a study where the confidence limits demonstrated that the trial was needlessly large [321].

Pocock and Hughes [322] have decided, on balance, that journals should have a consistent policy and that the 95% confidence intervals should be universally adopted rather than say 99% or 90% CIs. These authors also address the problems of determining the CI after early stopping of a trial and the CI when examining the results in a subgroup of the trial subjects.

SPLITTING OR SUBGROUP ANALYSES

Splitting the data, presumably into more homogeneous subsections, and then analysing these fractions is known as subgroup analysis. McMichael [323] stated, *"The aim of a statistical trial is to include all the unpredictable multitude of factors which can influence the outcome by a comprehensive sample. Unless the treatment shows a convincing difference in outcome for the whole group it is not permissible to separate out afterwards a sub-division of better results. Any subdivisions should be done on other criteria before the trial begins."* Bradford Hill would agree with the dangers of identifying a subdivision of better results, but when there are factors that influence outcome, he stated *"Surely it is our job and duty, to see whether in the analysis we can identify them and thus make them predictable."* He added, *"It is*

better to have looked and lost than never looked at all."
The first stage of any analysis must be to examine the treatment groups, as randomised and without any exclusions. This has been termed an analysis on the *intention-to-treat* principle. The next stage of the analysis may be to allow exclusions and examine groups who followed the protocol, an analysis on the on-randomised treatment basis. Lastly, the data can be analysed in prognostic and other groupings.

Analysis on the intention-to-treat and on-randomised treatment principles

The intention-to-treat principle
An analysis on the intention-to-treat (ITT) principle tests the strategy of offering a certain treatment to a group of subjects irrespective of whether they receive or persevere with the treatment. It is the safest way to analyse the results of a clinical trial in order to avoid bias.

The ITT analysis includes all subjects, is obviously not a subgroup analysis, is not subject to subtle biases, but has a major drawback in not accurately representing the results of continuous treatment. It is, however, the most relevant analysis for a clinician deciding whether or not to advise a particular treatment in the first instance. The on-randomised treatment or per protocol analysis is one confined to patients who complete the trial and take the prescribed treatment throughout. Since the patients selected for this analysis are usually those who have not defaulted, who have received the treatment, and who are thought to be "compliant" (ie, taking the treatment), this analysis is known as an "on-randomised treatment" (ORT) analysis.

Figure 14-4 illustrates a theoretical trial of active antihypertensive treatment versus placebo in patients over the age of 80 in which patients with compromised myocardial function cannot tolerate the postural hypotensive effect of the active drug and withdraw. Total mortality is the end-point of the trial. The ORT analysis compared the outcome in "completers" on active treatments with "completers" on placebo. A

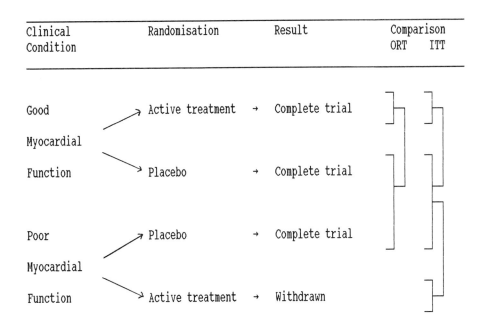

Clinical Condition	Randomisation	Result	Comparison ORT ITT
Good Myocardial Function	Active treatment →	Complete trial	
	Placebo →	Complete trial	
Poor Myocardial Function	Placebo →	Complete trial	
	Active treatment →	Withdrawn	

Figure 14-4. Theoretical comparison of antihypertensive treatment and placebo in hypertensive patients over the age of 80. ORT = on-randomised treatment analysis where some subjects with poor myocardial function and hypotension with active treatment are withdrawn. ITT = intention-to-treat analysis where subjects with a poor prognosis are not removed from the actively treated group.

"completer" is a patient who continues with treatment until the end of the trial or until they die. In this theoretical example, mortality will be artificially low on active treatment in an ORT analysis. This is because those patients who withdraw on active treatment may be expected to have a higher than average mortality in view of their compromised myocardial function. This is a biased subgroup analysis. The ITT analysis is not subject to such bias.

The on-randomised treatment (ORT) analysis
The ITT analysis does not reflect the effect of continuously taking the treatment. For example:

1.   In a small trial, default, dropout, or noncompliance may affect only one or two patients but their results may markedly

dilute the result of the treatment.

2. In a long-term trial of an active treatment versus placebo treatment, some patients who are intended to receive active treatment may not receive it and some who are intended to receive placebo will receive active treatment. Analysis on the intention-to-treat principle dilutes the effect of treatment.

In view of these problems analyses should be performed both on the intention-to-treat and on the ORT basis. An analysis on the ORT basis only considers those in the trial groups who received the treatments as specified in the protocol. Patients who do not adhere to the protocol are excluded and not transferred between groups. The worrying effects of such a selection have been discussed in chapter 7 and one trial that has been criticised for concentrating too heavily on this approach is discussed in chapter 19. In this trial it was decided that all events between randomisation and seven days could be ignored as an active treatment only exerts its effect after seven days of treatment [65]. Similarly, events were included for seven days after stopping treatment but not thereafter as the treatment continues to act for only seven days. The results of the trial have not been widely accepted owing largely to the selection of patients and their events for analysis.

Examination of the data in prognostic and other groupings

It is essential for the full analysis of a trial that the data are examined to determine the characteristics of patients in whom the treatment was most effective and least effective. This can only be done for large trials where different subgroups contain considerable numbers of patients. However, as Pocock and Hughes have pointed out [322] *"subgroup analyses are just one example of exploratory data analysis ... treatment differences should be viewed with great caution and ideally need validating in a subsequent trial"*. With this caveat it can be suggested that subgroup analyses are performed when [324]:

1. the analysis was intended at the design stage
2. the overall result of the trial showed a statistically significant result
3. the subgroup analysis is based on known biological differences between groups
4. the subgroup is not subject to an overt bias (such as ORT analysis)
5. a test of heterogeneity between strata is statistically significant

At the design stage the subjects may have been stratified prior to randomisation, for example as men or women, with the intention of analysing them separately. Conversely a subgroup analysis should not be excluded on the basis of not being based on prior stratification or specification. However the data should not be dredged in a 'negative' trial as some subgroups are bound to have fared better than others. An example of a biologically implausible subgroup analysis has been provided by Collins and his colleagues [325]. They performed a subgroup analysis by astrological birth sign in a trial of treatment in acute myocardial infarction [201]. Only patients born between October 24 and November 22 (under the sign of Scorpio) benefited from treatment. This trial also showed a significant benefit overall but not for the other astrological birth signs.

In a subgroup analysis the authors should state why they chose a certain subdivision for analysis and they may be able to report the results with more confidence if they stated their intention in the trial protocol. An interesting example was provided by the Hypertension Detection and Follow-up Program (HDFP) trial where patients in the United States of America were randomised to specialist care for their hypertension or referred to the usual community health services for treatment of their hypertension. The overall intention-to-treat analysis showed a benefit from specialist care [138] but when the results were broken down by sex and race, there was no observable benefit for white women [282]. This is an example of a reasonable subdivision of the data yielding results that may throw further light on the conclusions to be drawn from the trial. However, we are as yet uncertain as to what these conclusions should be.

The statistical test for differences in the effect of treatment between strata relies on the detection of a treatment-subgroup interaction term. Pocock and Hughes report that this was statistically significant in a trial of diltiazem versus placebo after myocardial infarction, when the subgroups were pulmonary congestion and no pulmonary congestion [322,326]. However, the overall result of the trial was not statistically significant and twelve subgroup analyses were performed. Allowing for the multiple analyses and employing a Bayesien approach to the problem Durrleman and Simon considered that diltiazem was not appropriate for either subgroup [327].

One large trial of treating hypertension in the elderly, the Medical Research Council (MRC) trial of treatment of hypertension in older adults [328] was interesting as subjects on active treatment were randomised either to the beta-blocker atenolol or to the diuretic combination of hydrochlorothiazide and amiloride. There was contamination of the groups but for 55% of the time those randomised to atenolol were known to be getting this drug alone and similarly 68% of the time the diuretic group were receiving this drug only. This trial therefore provided a definite opportunity for a subgroup analysis. The interaction term was of borderline significance and the results are summarised in table 14-4. The subgroup analysis suggests that diuretic treatment was to be preferred to beta-blockers in the elderly hypertensive.

Other trials have shown that beta-blockers can be given safely with a diuretic in the elderly [182,329, 330], nevertheless the MRC trial attempted to give this treatment on its own, with poor results. Moreover a beta-blocker was less successful than a diuretic in the MRC trial of younger hypertensives in a subgroup of male smokers, giving some consistency to the results [181]. The possible adverse effects of a

	Diuretic	Atenolol	Placebo
Number	1081	1102	2213

Fatal events per 1000 patient years

	Diuretic	Atenolol	Placebo
Stroke	2.5	3.3	3.3
Coronary	5.2*	8.2	8.6
Total	21.3	26.4	24.7

Treatment effect (%) compared with placebo

Stroke mortality
	Diuretic	Atenolol	Placebo
In trial	-24%	0%	-
Meta-analyses[+]	-36%,-35%,-33%		

Cardiac mortality
	Diuretic	Atenolol	Placebo
In trial	-40%	-5%	-
Meta-analyses[+]	-25%,-21%,-26%		

Total mortality
	Diuretic	Atenolol	Placebo
In trial	-14%	+7%	-
Meta-analyses[+]	-12%,-14%,-9%		

Table 14-4. Results of the Medical Research Council trial in the Elderly [328] and results of 3 meta-analyses based on all active treatments versus controls.
+ based on references 331, 332, and 333, respectively.
* 19% lower on diuretic (95% CI -2% to +36%)

beta-blocker in the elderly trial was also only observed in smokers. Those who consider that the adverse effects of a beta-blocker in the elderly may be biologically plausible believe the results of this particular subgroup analysis. Those who reject any plausibility do not. Table 14-4 also shows how different the results with a beta-blocker were from the conclusions of three meta-analyses of all treatments and trials in the elderly [331-333]. We do not know whether the result of the subgroup analysis should be accepted or not and it is unlikely that a new trial will be started using beta-blockers as first line treatment in the elderly.

AMALGAMATING TRIAL RESULTS - OVERVIEW OR META-ANALYSIS

The reason for doing a meta-analysis of randomised controlled trials is that individually the trials are too small to give reliable answers [334,335]. Furberg and Morgan [336] have defined six reasons for conducting such an analysis: to obtain a more stable estimate of the effect of treatment; to examine variability between the trials and assess the generalisability of the results; to perform subgroup analyses; to strengthen submissions on drug efficacy to licensing bodies; to identify the need for major trials; and to put any one trial result into perspective by examining all similar trials.

When may a meta-analysis be performed?

Only randomised controlled trials should be subjected to an overview [337-339], these trials should address a standard question (for example, surgery versus no surgery); the outcome of interest should be clearly reported; an intention-to-treat analysis should have been performed; the patients should be comparable (for example, in a certain age range); their diagnosis should be closely defined; the intervention and control treatments should be the same in each trial; and the trial should have been published in full. The advantage of complete publication is that the authors will have analysed their data fully and the article will have been subjected to peer review. Some workers, however, argue that the results should be included whether published or not [293], in order to avoid "publication bias", as results that do not reach statistical significance tend to remain unpublished.

Most meta-analyses include all trials irrespective of country where performed and the size of the trial. However, when examining an outcome that requires large trials some workers feel justified in excluding small trials that they judge ill conceived and unimportant [340].

The main difficulties encountered by meta-analysts are as follows [339]:

1. Which trials to include:
   All, avoiding (a) publication bias and (b) overviewer bias? [336]
   Those of good quality asking the same basic questions? [335]
2. Is there uniform patient selection? [341]
3. Are the treatments and patients comparable, or are you combining "apples and oranges and the occasional lemon"? [336]
4. Are definitions of outcome consistent? [336]
5. Problem of multiplicity - multiple looks, multiple subgroups, multiple methods of analysis, multiple outcomes. [336]
6. When to measure outcome? The length of follow-up is important if there are time-dependent treatment effects. [293,336]

Demets [342] concluded, *"while methods exist to combine studies, adequate procedures for deciding what to combine are not yet adequately developed ... Pooling or combining studies after the fact must never be viewed as a substitute for simple, well designed clinical trials"*.

How should a meta-analysis be performed?

The method of meta-analysis is important. It requires a thorough search to ensure that the appropriate trials are included. In particular we must take care to avoid selection of papers that support one view. A worrying trend is meta-analysis of trials selected, for example, on site of performance such as specialist clinics, general medical clinics, and general practices [343]. We must consider whether the treatment works overall.

The overview may consider a proportion of the trials or a subgroup within each trial. Demets [342] discussed combining data across trials into a single set and including stratification in the analysis. Peto and colleagues have argued strongly [293,335] that in each trial the expected number of events (E) should be calculated together with the difference between this figure and that observed (O-E). Peto et al's opinion is that physicians understand these calculations more easily than the logistic regression, maximum likelihood, and Cox regression techniques available for the same purposes. The O-Es are combined and the odds ratios are calculated. The Mantel-Haenzel method can achieve the same objective [344].

Too much time should not be spent on overviews. Three overviews in the elderly treated for hypertension all yielded similar results [331-333]. On the other hand, the first meta-analysis on the use of magnesium after a myocardial infarction considered 1300 patients and suggested a 50% reduction in mortality [345,346]. This was confirmed by two later separate meta-analyses [347,348]. However, a further large trial, LIMIT 2, included 2300 patients and found a reduction in mortality of 24% [349], and a megatrial, ISIS-4, included 58,000 patients [350] and found no benefit from magnesium. The Drugs and Therapeutics Bulletin reported this information and the fact that its consultants were unable to reach a consensus on whether magnesium should be employed or not [345]. The disagreement rested on whether or not the ISIS-4 trial would have shown a benefit if magnesium *had been given earlier*.

CONCLUSIONS

This chapter has considered the common errors to be found in the analysis of trial results, including errors concerning the experimental unit, failure to employ a statistical test, the use of an unspecified or incorrect test, the use of one-tailed tests, and the incorrect interpretation of a negative result. Also outlined were analyses to demonstrate that randomisation has been effective and tests of significance given for both quantitative and qualitative data. It is essential that the data be checked for errors, the distribution of the data examined, and a transformation performed where necessary. The calculation of confidence limits was discussed and their importance stressed. The analysis of survival data was briefly reviewed and the analysis of data on the intention-to-treat and on randomised treatment principles was discussed. Examples of splitting the data and analysing subgroups were presented, and lastly, the amalgamation of the results from different trials in a meta-analysis was reviewed.

Subgroup analyses have to be treated with caution but meta-analyses are probably more reliable. Although Wittes [341] concluded *"For individual decision-making ... it seems unlikely to me that data from overviews will-ever be decisive"*, properly conducted overviews have a valuable place in research methodology.

15. THE EVALUATION OF SUBJECTIVE WELL-BEING AND MEASUREMENT OF QUALITY OF LIFE

INTRODUCTION

Trials of treatments have, of necessity, to concentrate on a few major end-points. Thus a very large trial may be concerned solely with survival, recording fact of death, date of death and cause of death. Nevertheless the results of such a trial have to be evaluated in conjunction with information in the adverse effects of treatment to give the benefit : risk ratio for the treatment. The detection and quantification of severe adverse drug reactions (ADRs) is discussed in chapter 16 and the measurement of the benefit : risk ratio in chapter 17. With dramatic life saving treatment, such as drugs employed to induce remission in acute leukaemia, assessment of severe ADRs may suffice, the assumption being that an acceptable quality of life (QOL) will be enjoyed during remission. However, the QOL may not be acceptable and in preventive medicine, the gains for the average individual may be small. Only a few of the subjects may be expected to get the disease and in only a proportion of these may the preventive medicine be effective; yet adverse effects may affect all the subjects. In such a situation small risks may outweigh small gains and it is important to measure subjective well-being and QOL.

Evaluation of subjective well-being

The subjective well-being of a patient is evaluated by the identification and quantification of symptoms. If due to treatment they have been termed 'symptom side effects' [351]. These are usually subjective with no objective evidence, for example a headache, but they may also be associated with physical signs such as flushing or swelling of the ankles.

Variables such as age, gender and psychological state are often strongly associated with symptoms and need to be assessed.

The recording of a symptom as present will depend on the recognition of the symptom by the patient; the reporting of the symptom by the patient; and the recording of the symptom by the observer.

Anxious or depressed patients often have a multitude of complaints. Some of these may have an organic basis but would remain unnoticed when the patient is well, or when recognised, would soon be forgotten. In order for a patient to report a symptom it must be perceived and then remembered.

When a patient has recalled the presence of a symptom he or she still has to report the fact to the observer. Both the patient's attitude to the observer and the observer's relationship to the patient may influence whether a symptom is reported or not. For example, if a patient is very grateful to a physician for the care he has received he may be unwilling to report symptoms, viewing such complaints as an expression of ingratitude. On the other hand, a patient with a grievance may possibly list more symptoms.

The attitude of the observer is often crucial in the reporting of symptoms. A sympathetic observer may appear relaxed and encourage conversation in the form of initial small-talk that reassures the patient and leads to a fuller disclosure of complaints. Also the observer may assiduously enquire whether or not the patient has certain symptoms and this may lead to a high rate of reporting. The formidable, brusque observer is unlikely to be told of so many symptoms.

Patients report subjective assessments of their well-being that cannot be confirmed and different observers may interpret this information in varying ways and record invalid and unrepeatable data. The observer must report the patients' symptoms without bias. In a clinical trial a physician may take little notice of a symptom that appears to be of a psychological origin, especially if numerous such complaints are reported at the same time. The

variation in the frequency of reporting symptom
side effects in drug trials is illustrated in
table 15-1.

Table 15-1. The percentage of hypertensive patients complaining of sedation when
taking methyldopa; symptoms assessed by interviewer.

% complaining on methyldopa = (1)	Control treatment	% complaining on control treatment = (2)	% complaining due to methyldopa = (1) - (2)	Direct question (DQ) or spontaneous report (SR)	Reference
10	Clonidine	56	?	DQ	352
15	Diuretic/ Placebo	6	9	?	353
24	Bethanidine	12	12	SR	354
25	Debrisoquine	18	7	DQ	355
37	Clonidine	63	?	DQ	356
47	Guanethidine	0.5	46	DQ	357
75	Guanethidine/ Bethanidine	13	62	SR	358
83	As above	51	32	DQ	358

The proportion of patients who reported
sleepiness in various trials of the
antihypertensive but sedative drug, methyldopa,
varied between ten percent and 83 percent. Such
variability is obviously not acceptable but does
occur with current methods of assessment. Table
15-1 also gives details of the control treatment
and the percentage complaining of sedation on
this treatment. The difference between the
proportion complaining in the methyldopa group
and the proportion in the control group provides
an estimate of the percentage of complaints
attributable to methyldopa. The attributable
percentage varied from seven percent to 62
percent and did not appear to be affected by

whether the symptom was elicited by direct questioning or arose as a spontaneous report. However, with such variability it is important to standardise the method of collecting these data. One treatment, clonidine, produced sedation more often than methyldopa and could not be employed as a control but the other control treatments were not known to produce sedation.

The recording of symptoms in clinical trials is difficult but of great importance. The observer may simply record any complaints reported to him, enquire specifically about certain symptoms, or ask the patient to complete a questionnaire without further assistance. The first method can be termed recording of spontaneous complaints; the second, interviewer-administered questioning; and the third, self-administered questioning. The questionnaire methods rely heavily on asking a sensible question. This chapter will discuss standardised methods for eliciting such data. Trials of treatment in psychiatric patients will also be reviewed. Lastly, the importance of measuring the overall quality of life will be considered and methods for quantifying this concept will be discussed. We shall be concerned primarily with recording symptoms and other data of a sensitive and subjective nature. However, the methods and concepts are valid for obtaining any verbal information from a subject.

SPONTANEOUS REPORTING OF SYMPTOMS

When a patient reports a symptom without prompting, this suggests that the complaint is troubling the patient and is a more severe problem than one detected only by direct questioning. It is therefore important to record such information before proceeding to ask specific questions. The difficulty is in standardising the procedure from one patient to the next and from one occasion to another. The observers should be asked to follow a consistent protocol, for example:

Step one   Ask the patient to sit down, "Good

morning (afternoon). Please sit down."

Step two   Ask the patient, "Have you any problems?" This will have to be modified if the patient is known to be suffering from a complaint. The question would then be, "Have you any problems apart from ...?"

Step three   The observer must carefully record all the symptoms reported by the patient.

Step four   Code and analyse the information. This may be very difficult owing to the multitude of possible answers and to error or bias in those responsible for coding [359]. In a small trial the responses can be listed and only those analysed that are known to be of interest or reported with any frequency.

Having ascertained spontaneous reports of symptoms, the observer may proceed to ask questions. We must consider what constitutes a good question.

A GOOD QUESTION

A good question must be easily understood by the subject, limited to an exact period of time, have a high repeatability and good validity when used to detect a clinical condition, have the same meaning for different individuals (be unambiguous), concern only one item at a time, be grammatically correct, and result in an answer. A question may also be open, with the respondent answering in his own words, or closed, where he selects from alternative answers. Any tendency on the part of respondents to always agree either to the first answer or to a positive or negative statement must be discouraged. Certain characteristics of good questions were also reviewed in chapter 10.

i) The question must be clear
Medical personnel often overestimate their patients' understanding of medical terms and no

ambiguous terms should appear in the question. Trials will include many subjects of lower than average intelligence and the questions should be designed so that these persons can understand and answer. Long words and double negatives should be avoided and whenever possible a difficult long word should be replaced by a short one. Bennett and Ritchie [209] have reviewed the qualities of good questions and considered that leading questions should be avoided where possible and that vague terms such as *occasionally* or *often* must be replaced by precise numerical terms. It is also important to limit the time over which the symptom should be recalled.

Consider the following questions:

"Since your last visit have you often felt sleepy during the day?"
Answers: Yes/No
"Have you, in the last three months, noticed weakness in the limbs?"
Answers: Yes/No

The first question uses the term *often* and also leaves the duration vague; however, it would be suitable in a trial with a short fixed interval between visits. The second question could be used but subjects are unlikely to recall symptoms occurring 3 months ago, and a shorter duration, say 2 weeks, may be more appropriate.

ii)   The question must have a high repeatability
The measurement of repeatability has been discussed in chapter 13 and indices have been provided that can be applied to the questionnaire answers. To assess repeatability, a group of persons should be asked the same question twice, with the occasions separated by a short interval, say, two weeks. The subjects answering the questions should be from a similar background to those who will be recruited to the trial. The questions must not be repeated after a very short interval, lest the subjects remember their first replies, and a number of questions can be tested simultaneously in order to limit this recall. A question that is highly repeatable may be well understood but not necessarily valid: the question may not measure what is intended.

iii) The question must be valid
Validity has been discussed in chapter 13 and can be defined as the extent to which the question measures what it is supposed to measure. An example can be provided by a series of questions that were designed to detect the symptoms that occur with a marked fall in blood pressure. Certain antihypertensive drugs produce these symptoms when the subject stands up. Three of the questions were as follows [206]:

"Since your last visit have you suffered from unsteadiness, light-headiness or faintness?"
Answers: Yes/No
"Does the unsteadiness or faintness occur only when you are standing?"
Answers: Yes/No
"For how many hours in the day are you troubled by unsteadiness or faintness?"
Answers: Less than one hour/one-two hours/more than two hours

When the blood pressure of the respondents was measured standing and lying, a larger than average fall in pressure on standing was observed when positive answers to the first two questions were reported with a duration of less than one hour. Thus, this series of questions had a degree of validity provided the answer sequence was yes/yes/less than one hour. One question, designed to identify a previous stroke was not successful. The question was:

"Have you ever had, without warning, sudden loss of power in an arm?"
Answers: Yes/No

Most of the patients who responded positively to this question had not had a cerebrovascular accident at any time. The false positive rate was therefore high and this question was not acceptably valid. Perhaps the question, *"Have you ever had a stroke?"* would be more suitable. However, this direct question may still have a high false positive rate owing to confusion with other episodes of illness and a high false negative rate if minor cerebrovascular accidents are to be detected. Medical terms are frequently not understood [360] but may be better understood by those who have the condition that is to be detected [361]. In general, medical terms should be avoided.

iv)   The question must not be ambiguous
A question was designed to detect the presence of diarrhoea and was phrased as follows:

"Are your motions often loose or liquid?"
Answers: Yes/No

Faeces are often referred to motions in colloquial English, but this attempt to use simple words may have led to ambiguity. The question could be interpreted as an enquiry about physical mobility and should be modified to prevent this possible error.

v)   The question should only make one enquiry
The first question in section iii) above asks whether the patients have had *unsteadiness, light-headiness or faintness*. Patients answered yes if they had suffered from any one of these symptoms or indeed all three. It would be preferable to ask about each symptom separately as it is possible that faintness results from postural hypotension (the condition that is to be detected) whereas unsteadiness may be due to arthritis of the legs or vertigo. If the latter is true, the inclusion of the symptom unsteadiness in the question may increase the false positive rate.

vi)   The question should be grammatically correct
It is more important that the question is well understood by the man in the street than whether or not it is grammatically correct. A standardised, repeatable, and widely used questionnaire concerning angina [362] was criticised on grammatical grounds by London Civil Servants who were planning to administer it. However, the extensive use of the questionnaire in other studies precluded any changes, as comparisons with other populations would then be

impossible and further studies on repeatability
and validity would have to be carried out on the
grammatically improved questionnaire.

vii) The question must be answered
When a question produces embarrassment or offence
it may not be answered, even with full guarantees
of confidentiality. For example, it may be
important to ask questions concerning sexual
activity, but older subjects, whether sexually
active or not, may be less willing to answer than
young patients. Similarly, religious beliefs may
prevent a question being answered. A Muslim may
be as embarrassed if asked whether he drinks
alcohol as a Christian would be if asked if he
commits adultery.

A good question may be open-ended or closed.
An open question is one where the respondent
writes his or her answer or an interviewer
records the exact reply. With a closed question,
the respondent chooses between the answer options
provided. The question *"Have you any problems?"* is an
open question and the questions in section iii)
above are examples of closed questions. With an
open question the subject has to recall something
whereas with a closed question he is asked to
agree with a statement. This may constitute a
fundamental difference between the two varieties
of question [363].

Open questions should be used during the
initial stages of questionnaire design even when
closed questions are finally intended [209].
Mellner considered that the failure to employ
open questions may lead to a loss of information
[364], but Belson and Duncan suggested that more
information may be derived from closed questions
[365]. Closed questions certainly minimise the
problem of recall and are most suitable when
there are a limited number of possible responses
[366]. Responses to open questions may be very
difficult to code and analyse [359].

When asking a series of questions the
tendency for a respondent to select a particular
answer or to give a certain reply [367] must be
discouraged. This is known as a response set when
an individual tends to select a particular answer
whether correct or not: for example, the first

answer, the positive response, the negative answer, or the neutral response such as *"I don't know"*. The effect of any response set may be limited by continually varying the position of any answer options [368] and by providing a greater variety of answers than yes and no [369]. If the position of an answer is only varied occasionally, then errors may occur. For example, six yes/no options followed by a no/yes option may lead to an error in completion.

The problem of a response set may tend to be less with open questions and an interviewer, than with closed questions and a self-administered questionnaire.

THE INTERVIEWER-ADMINISTERED QUESTIONNAIRE

There are certain advantages and disadvantages of having the questions read by an interviewer rather than having the respondent complete a self-administered questionnaire (table 15-2). The interviewer must be taught to standardise his or her interview so that a question is always asked in the same manner, in the same order, and after an identical introduction. This training may be difficult and does not ensure that two interviewers will get the same responses. However, interviewers have certain advantages. They will have an impression of whether or not the question is understood and can make additional information available if required. Extra clarifying statements must be stipulated in advance and printed on an interviewer's form. With an interviewer, the subjects will also be able to answer the questions even of they cannot read or have forgotten their glasses. In addition, when a series of questions has to be asked as a consequence of an initial (usually positive) response, the questions are easier to ask using a trained observer, as complicated instructions have to be added to a self-administered questionnaire. For example, the instruction, *"If the answer is 'No' please proceed to question ..."* may confuse a respondent. Finally an interviewer can ensure that an answer is obtained

Table 15-2. The advantages and disadvantages of interviewer and self-administered questionnaires.

SELF-ADMINISTERED	INTERVIEWER-ADMINISTERED
Disadvantages	
Excludes more patients	Difficulty in training
More missing information	Race, sex and age of interviewer may influence results
Complexity of branching questionnaire	Difficulty in standardisation
Advantages	
Reproducible	Subsidiary information may be provided
Less expensive	Includes most patients
May be less embarrassment	Most data complete
	Can probe to elicit unexpected information

to nearly every question, whereas with a self-administered questionnaire subjects may not answer all the questions. However, the use of an interviewer is expensive and the results subject to observer variation despite rigorous training.

THE SELF-ADMINISTERED QUESTIONNAIRE

The use of a self-administered questionnaire removes the effect of observer variation. The method also tends to be much less expensive as the questionnaire has only to be handed or posted to the subject and an interviewer does not have to be trained and employed. However, the subjects have to be able to read, and if necessary, have their reading glasses available when given the self-administered questionnaire. They may fail to complete some of the questionnaire. Where possible, the self-administered questionnaire should be administered under standardised conditions: for example, in a quiet, private room prior to a clinical investigation.

In a comparison of results from an interviewer and self-administered questionnaire it was found that the self-administered questionnaire gave a higher proportion of positive responses to sensitive questions [206]. For example, male patients were asked about impotence by a male observer, but a self-administered questionnaire including the same questions suggested a higher rate for this complaint (47 percent against 28 percent). It appeared that the patients were reluctant to admit so readily to this embarrassing symptom when asked by an interviewer. In this study, less sensitive questions were not affected by the method of collecting the information. With very personal questions, the self-administered questionnaire may have an advantage.

The self-administered questionnaire is also particularly useful in multicentre international trials where interviewer training and standardised conditions would be difficult to achieve. When there are differences in language between the centres, the questions, whether self-administered or not, must be translated from the original language into the new and then translated back into the first language by someone who has never seen the original questions. If the back translation of a question is not close to the original, the first translator must try again until satisfactory matching is achieved.

Whether a questionnaire is self-administered or not, the origin and purpose of the questionnaire must be fully explained to the respondent and the questions must follow a logical sequence. It must be remembered that the respondent will expect some relationship between adjacent questions; the order of administration may also affect the responses [370].

RANDOMISED CONTROLLED TRIALS IN PSYCHIATRIC PATIENTS

Randomised controlled trials in psychiatric patients are often concerned with subjective changes such as symptoms. In psychiatry, diagnostic difficulties are more marked and informed consent may be difficult to obtain. The response to treatment may be difficult to measure, treatment may have to be prolonged to exert an effect, habituation may occur to drug treatment, a variable dose may be required, and the drugs may produce many side effects.

The assessment of new drugs for use in psychiatry

A new drug for use in cardiovascular medicine may have demonstrable effects in animals and human volunteers and the clinical efficacy of this type of drug can be studied in animal models of disease, but animal models of schizophrenia, depression, and anxiety are less well developed. Early in their development, drugs for use in psychiatry have to be tried on patients.

Diagnostic difficulties in psychiatry

Defining a psychiatric condition may present great difficulty. A Medical Research Council Trial of treatments for depression employed an operational definition based on clinical impression, the presence of certain symptoms, a short duration of illness, and lack of previous treatment [371]. However, international differences on whether a mentally disturbed patient has schizophrenia, manic depression, or another condition exist so that it is very important to carefully define the type of patient who may enter a trial.

Obtaining informed consent from the patient

This problem arises mostly with schizophrenia, severe depression, learning difficulty and dementia. If informed consent cannot be obtained from the patient, it must be obtained from the closest relative or guardian.

Measurement of response

Hamilton [372] has defined four categories of improvement: subjective changes; objective changes; improvement in personal relations; and greater working capacity.

i)   Subjective changes

Symptomatic improvement may be determined by an interviewer or a self-administered questionnaire. When a self-administered questionnaire has been used in psychiatry to diagnose or quantify the degree of depression or anxiety, the questionnaires have been termed *self-rating scales*. Discussion of these scales is beyond the scope of this book and not all are useful for detecting the response to therapy that may be observed in a trial. For example, the well-known Eysenck Personality Inventory [373] measures whether or not the subject has a neurotic personality. This trait may be constant and not susceptible to short-term fluctuations.

Hamilton has reviewed the self-rating scales that may be employed to assess the state of anxiety. He found that at least one self-administered questionnaire can detect drug effects, namely, Taylor's Manifest Anxiety Scale [374]. Hamilton also considered that three interviewer-administered scales can be useful: the Brief Psychiatric Rating Scale [375] for all psychiatric symptoms; the Symptom Rating Test [376]; and Hamilton's rating scale for anxiety states [377].

A variety of scales are also available for the assessment of depression. These rely on symptoms and care has to be taken that a symptom employed in the scale, such as early waking, is not also a side effect of the drug being tested. If this is true the scale will alter, not because of a change in the degree of depression but from a direct drug effect. Some questionnaires are completed by an observer, such as the Montgomery-Asberg Depression Rating Scale [378], and some by the patient, for example the Beck Depression Scale [379].

Outcome in schizophrenia can be estimated from symptoms, relapses and returns to hospitals.

Leisure activities, domestic relationships, behaviourial disturbance and other aspects of social functioning have been assessed using the Social Performance Schedule [380].

A trial in psychiatry must employ standardised methods of assessment, and one or more rating scales should be used where appropriate. The greater the difficulty in assessing a response, the greater the importance of using standardised methods that can be reproduced by another investigator.

ii) Changes in personal relationships

The ability to resume full-time work could be an important end point of a trial as could discharge from hospital, discharge from care, readmission to hospital and, in depression, the frequency with which electroconvulsive treatment has to be employed.

Any change in behaviour that can be documented must be carefully defined at the onset of the trial. Hamilton [372] provided two examples: outbursts of temper that could be observed in a hospital ward and the frequency of going out of doors in a patient with a fear of open spaces (agoraphobia).

Problems of drug trials in psychiatric patients

A long duration of treatment may be necessary, the dose of drug may be varied, habituation may occur, and side effects may prove very troublesome. Shepherd [381] considered that the highly lipid soluble drugs that act on the brain may be metabolised at more variable rates than other drugs, making a variable dose schedule of greater importance. Habituation to sedative treatment often results in the side effect of sleepiness being lost after a week or so and Hamilton suggested starting with very small doses and increasing slowly, even though this will prolong a trial [372]. He also pointed out that a patient habituated to one drug may possibly not respond to a second and cautioned against the use of cross-over trials in psychiatry.

The difficulty of conducting trials in psychiatry should not inhibit such investigations as even greater problems may arise from a failure to perform randomised controlled trials. The randomised controlled trial is essential in the field of psychiatric investigation.

THE IMPORTANCE OF MEASURING THE QUALITY OF LIFE DURING A TRIAL

Treatment may produce side effects that interfere with a variety of aspects of the quality of life. For example, antihypertensive medication can produce gastrointestinal effects that interfere with the enjoyment of food and drink, and side effects on the cardiovascular system may reduce sporting activities. Similarly, pharmacological effects may prevent the enjoyment of sex, sedation may interfere with both work and play, and some drugs may produce depression and interfere with personal relationships and social contacts. It is not sufficient to demonstrate that an antihypertensive drug lowers blood pressure. It has also to be proved that side effects are not severe and that the quality of life does not deteriorate. It is to be hoped that a treatment does not impair general well-being and that an adverse effect can be proven absent.

Bullinger [382] has defined QOL and her definition may be briefly summarised as *'A multi-dimensional construct of physical, psychological and social aspects of well-being and function as perceived by patients and observers'*.

In this section we will consider methods that have been employed in measuring QOL and have (or are thought to have) the following characteristics.

i)   Relevance
The questionnaire covers areas of concern to the patient group. For example an assessment of performance at work would be important for a diabetic aged 40 years but not a patient with chronic heart failure aged 80.

ii)   Validity and repeatability
These are discussed in chapter 13.

iii)   Acceptability
The information should not take too long to collect, the content of the questionnaire should not upset the respondent and the completion rate should be high.

iv)   Sensitivity to change
The method should produce results that change with active treatment.

v)   Ease of analysis
The method should be easy to code and score, sometimes a problem with free text in unstructured interviews.

Should a Profile or an Index be employed to measure QOL?

An instrument that measures QOL may be termed either a profile or an index. A Profile assesses different components of QOL and 'profiles' them as does a battery of tests. An Index should provide a single overall score. In fact many questionnaires are named an Index when they are a Profile. Some instruments provide both a Profile and an Index.

General measures of QOL

Three general measures are listed in table 15-3, the Sickness Impact Profile [383], the Nottingham Health Profile [384] and the Quality of Well-Being Scale [385].

i)   The Sickness Impact Profile (SIP)
The SIP consists of 12 categories including physical mobility, work, emotional behaviour and social interaction. An assessment of pain is not included and 136 questions are asked, either self-administered or interviewer-administered.
    The SIP has been employed in several randomised controlled trials and proved to be sensitive to the effects of different treatments including early exercise and counselling for myocardial infarction patients [389], hyperthyroidism [390] and angina [391].

ii) *The Nottingham Health Profile (NHP)*

The NHP includes the same areas as the SIP but in only 45 questions. However the profile includes 8 questions on pain and one on sexual functioning. The NHP was sensitive to the benefits of a non-steroidal anti-inflammatory drug in the treatment of osteo-arthritis [392]. The profile may be self-administered or interviewer-administered.

iii) *The Quality of Well-Being Scale (QWB)*

Unlike the SIP and NHP, this scale is not a Profile but an Index. It is based on the work of Fanshel and Bush [393] that provided a numerical framework between 1 (perfect health) and 0 (dead) according to the level of morbidity (see below). Thirty five questions are pursued using a flowchart by a trained interviewer. The scale is sensitive to the benefit of oral gold (auranofin) in rheumatoid arthritis [394] and to the effect of zidovudine in AIDS [395].

Table 15-3.   Measures of Quality of Life (QOL)

General measure of QOL		Reference
1.	Sickness Impact Profile (SIP)	383
2.	Nottingham Health Profile (NHP)	384
3.	Quality of Well-Being Scale (QWB)	385

Psychological Dimension		
1.	Psychological General Well-Being Index (PGWB)	386
2.	Profile of Mood States (POMS)	387
3.	Hospital Anxiety and Depression Scale (HAD)	388
4.	Symptom Rating Test (SRT)	376

THE CHOICE OF A GENERAL MEASURE OF QOL

All three measures have their advantages and disadvantages. The SIP is comprehensive but takes 45 minutes to complete, may be self-administered or interviewer-administered and is sensitive to change. The NHP is much shorter and quicker to complete, includes a useful section on pain, should be first choice for painful conditions but has disadvantages compared with the SIP. The SIP, NHP and QWB were compared in 56 patients with angina prior to embarking on a randomised controlled trial [391,396]. Table 15-4 shows that all three instruments differentiated those with mild or no heart failure (New York Heart Association (NYHA) Grade I) from those with more severe disease. However the SIP and NHP were self-administered and the QWB scale required an interviewer. Moreover the total score with the

| | NYHA Classification | | | |
	I	II	III	P
Number	10	25	21	
SIP				
Physical score	5.4	13.1	19.3	0.01
Psychological score	8.1	15.8	25.1	0.02
NHP				
Physical mobility	5.4	17.2	24.9	0.03
Social isolation	0.0	5.4	12.4	0.02
QWB				
Physical activity	0.04	0.06	0.06	0.04
Symptom score	0.24	0.26	0.28	0.02
Total score	0.68	0.62	0.62	0.26
SRT				
Somatic score	1.9	3.4	4.5	0.06

Table 15-4. Measures of QOL in 3 New York Heart Association (NYHA) grades. SIP = Sickness Impact Profile, NHP = Nottingham Health Profile, QWB = Quality of Well-Being Scale and SRT = Symptom Rating Test. The statistical test examined for differences between the groups [396].

QWB instrument did not relate to the NYHA grade and the scale concentrated on the most distressing symptom, often breathlessness, to the exclusion of angina. The QWB also includes only 2 questions on psychological well-being. The QWB was therefore not considered for use in the final trial. Nine results were compared between the SIP and NHP. Not surprisingly, as the SIP is based on more information than the NHP, no median equalled zero with the SIP and 6 did so with the NHP. The potential for improvement was therefore less with the NHP. For the coefficients of variation, 8 of the 9 were also lower with the SIP. The SIP was therefore chosen for the final trial of *continuous* transdermal glyceryl trinitrate against placebo and detected an adverse treatment effect (probably due to headaches) in the active group shown by a deterioration in the psychosocial scores compared to placebo (Table 15.5).

METHODS ADDRESSING DIFFERENT AREAS OF QOL

The psychological dimension

Table 15-3 listed four measures of QOL that address psychological well-being and have been employed outside the field of psychiatry.

Benefits (none)

Attack rate not altered
Sublingual GTN consumption not altered

Risks

Withdrawal due to headache in 5%
Improvement in SIP physical score of 5% compared with an 11% improvement on placebo
Improvement in SIP psychosocial score of 8% compared with 17% improvement on placebo

---

Table 15-3. Benefit : risk comparison in 210 patients with angina treated continuously for 2 weeks with 5mg transdermal glyceryl trinitrate compared with 207 who received placebo [391]. Sublingual GTN = glyceryl trinitrate tablet absorbed under the tongue.

The Psychological General Well-Being Index (PGWB) [386], the Profile of Mood States (POMS) [387], and the Symptom Rating Test (SRT) [376] have all been employed in trials of treatment with antihypertensive drugs [397].

The PGWB and SRT have identified the ability of antihypertensive agents such as methyldopa and propranolol to cause depression, confirming clinical experience. The POMS has not been used in studies of propranolol and methyldopa, and the responsiveness of this instrument has not been determined. The SRT is based on symptoms and the POMS is based on mood. It is possible that SRT identifies symptoms that are related to the drug but may not necessarily be related to mood. Such an example would be 'feeling dizzy or faint' or 'feeling tired or having little energy'. If this is so, the POMS may have an advantage. However, at the present time most experience has been obtained with the PGWB and SRT.

The Hospital Anxiety and Depression Scale (HAD) has been recommended for use with cancer patients [398,399].

Specific measures for certain conditions

The Rotterdam Symptom Checklist [400,401] has been recommended for cancer patients [398,399]; a symptom questionnaire together with a Health Status Index (HSI) [402] derived from the work of Fanshel and Bush [393] for hypertensive subjects; and disease specific methods have also been discussed for asthma [403], skin disease [404] and rheumatoid arthritis [405].

CONCLUSIONS

In randomised controlled trials attention must not only be directed to objective measurements of outcome but also to the symptomatic well-being of the subjects. This chapter considered how the presence of symptoms should be determined from spontaneous reports and interviewer- and self-administered questionnaires. The characteristics of a good question were discussed including comprehension, validity, and repeatability. The importance and difficulty of performing randomised controlled trials of treatment in psychiatric patients, where the outcome is often subjective, is discussed. Finally, the assessment of both symptoms and other aspects of a patient's life are considered and how the quality of life may be measured. General measures of QOL and measures addressing a limited area of QOL are reported.

16. THE DETECTION OF ADVERSE DRUG REACTIONS

An adverse drug reaction (ADR) has been defined by the World Health Organisation [406] as *"one which is noxious, unintended and occurs at doses used in man for prophylaxis, diagnosis or therapy"*. This has been extended by *'excluding failure to accomplish the intended purpose'* [407] in order to prevent lack of efficacy being classified as an ADR [408]. During the course of a randomised trial it may not be recognised that an occurrence is indeed an ADR. Only the final analysis may reveal that the adverse event is due to a drug. This has led to the collection of Adverse Events (AEs) during a trial. These have been defined succinctly as a *'Particular untoward happening experienced by a patient undesirable either generally or in the context of the disease'* [409]. During a trial it is adverse events that are collected and they should include abnormal signs, symptoms and laboratory tests [408,410]. The United States Food and Drug Administration (FDA) definition includes *'or significant failure of expected pharmacological action'* [411]. Needless to say all hospital attendances or admissions must be included. The detection of symptom side effects was discussed in chapter 15 and in this chapter we shall consider only life-threatening events.

An ADR will be more easily detected when it is an event known to be associated with drug treatment and is also known to be relatively rare in the absence of such treatment.

IS THE EVENT KNOWN TO BE PRODUCED BY DRUGS?

Figure 16-1 gives the hypothetical steps in the recognition of an ADR. If a condition is suspected to be an adverse event and is uncommon in the absence of drug treatment, then an ADR will be detected. If an ADR mimics a common event, usually unassociated with drug treatment,

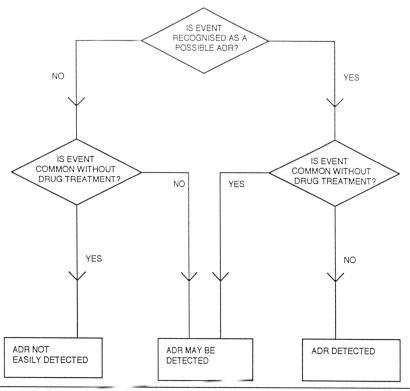

Figure 16-1. Flow chart to illustrate the detection of an adverse drug reaction (ADR) in a trial. Detection is greatest when an event is recognised as a possible ADR and is uncommon in the absence of drug treatment. An ADR may easily be missed when it is not known to be an adverse effect of treatment and is common in the absence of treatment. The large randomised controlled trial provides the best opportunity of detecting such an ADR.

it may still be detected, as will a previously unsuspected ADR if it is rare in the absence of treatment. The large randomised controlled trial gives the best opportunity for detecting previously unsuspected ADRs, provided data on all events are collected during the trial and subsequently analysed.

FREQUENCY OF THE ADVERSE EVENT

Figure 16-2 provides a schematic representation of the chance of detecting an adverse event during a trial and the alternative, of finding a

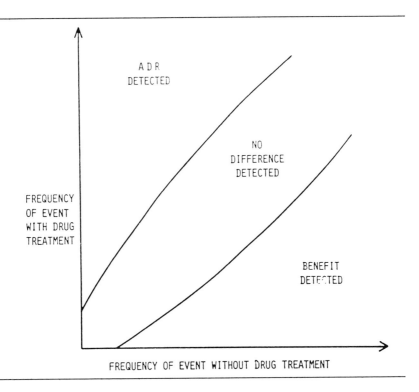

Figure 16-2. Schematic representation of the chance of detecting an adverse drug reaction (ADR) in a randomised controlled trial. When the ADR is common in the treated group and rare in the control group, an ADR may be detected. Conversely, an event that is less common in the treated group will represent a benefit of treatment. The central area represents the results when neither an advantage nor disadvantage of treatment is detected.

reduction in this event in the treatment group (a benefit from treatment). An ADR will be detected when the frequency of the event is low in the absence of a particular drug treatment and high when the drug treatment is employed. Conversely, if the drug frequency reduces the incidence of an event that is common without treatment, then a benefit may be detected.

The large randomised controlled trial is the ideal method for detecting ADRs which occur with a frequency greater than one in 300 treated patients and not in controls [412]. However, trials are unlikely to detect rare ADRs such as aplastic anaemia with phenylbutazone (one in 33,000) or thrombocytopenia with diuretics (one

in 15,000) [413]. There was also a notable
failure of one large trial to detect the more
common oculocutaneous syndrome with practolol
[414], possibly due to failure to collect the
appropriate information during the trial [409].
With very rare ADRs, the randomised controlled
trial is not the appropriate method for detecting
these conditions. It is to be hoped that clinical
intuition, specially designed surveillance
programmes, and the routine examination of vital
statistics will lead to the discovery of rare
events. Lewis has prepared a table of the number
of patients who would have to be surveyed in a
treated group according to the background
incidence of an event and the additional
incidence due to the drug. With a background
incidence of one in 1,000 and a doubling of this
incidence due to the drug, 32,000 patients would
have to be surveyed [415].

A small clinical trial including less than
50 patients stands little chance of detecting an
ADR with any certainty. However, small trials
have produced clues to the presence of adverse
drug reactions, and large trials have succeeded
where the ADR is a common condition. Most life-
threatening ADRs occur with a frequency of less
than one in 50, a possible exception being
provided by primary pulmonary hypertension with
aminorex fumarate.

THE SMALL TRIAL AND THE ADR

Adverse drug reactions may be detected in small
clinical trials, but the results are often
difficult to interpret. Vere discussed this
problem [416] and reviewed four trials of
chlorpromazine that reported in 1954 when the
first case reports of jaundice with this drug
were appearing. In these trials 17, 22, 24, and
27 patients were given chlorpromazine. The trial
with 27 patients did report one case of jaundice
but this could not be attributed definitely to
the drug. Similarly, in a trial of the
antihypertensive drug guanoxan, two of 160
patients developed obstructive jaundice. Seventy
percent of the 160 patients showed rises in serum

transaminase concentrations during treatment against forty percent before treatment [417]. The authors commented *"Although a rare direct toxic effect ... upon the liver cannot be excluded ... this explanation of the remaining abnormalities appears to be most unlikely ... no case of jaundice has been described which can, with certainty, be ascribed to therapy with these compounds."* All would now consider that the episodes of jaundice were due to the drug.

Trials can also provide evidence of serious toxicity without the end result of the toxic process being observed. For example, a small trial showed evidence of an overall reduction in peak expiratory flow rate with a drug known to produce bronchospasm in susceptible subjects [207]. Similarly, the fact that a drug produces hepatitis may be detected in a small trial in which liver function tests are monitored; and a tendency to produce serious marrow depression may be detected by a significant reduction in platelet or white cell count.

THE LARGE TRIAL AND THE ADR

Many methods of detecting ADRs depend on case reports, either published or incorporated in central registers. These reports tend to involve drugs that are already under suspicion and also to concentrate on events that have been previously shown to result from drugs – for example, aplastic anaemia and hepatitis. When an adverse effect mimics or is identical to a common disease, it is less likely to be recognised as an ADR. The clinical trial, with its integral control group, is capable of detecting an excess of common events when associated with drug treatment.

Table 16-1 lists some adverse drug effects that have been detected in three large controlled trials. In the Coronary Drug Project Research Group trial [62] there was a one in 50 excess of nonfatal myocardial reinfarctions in those given high-dose conjugated oestrogens for one year, an excess of arrhythmia in those given niacin for a year, and an excess of one in 500 in gallbladder disease in those given clofibrate [418]. In the

Table 16-1. Adverse drug reactions detected in three large controlled trials. The frequency represents the difference between the rate observed in the actively treated group and the appropriate control group.

Drug	ADR	Frequency	No. of patients in trial receiving drug	Reference
CONJUGATED OESTROGEN 5mg daily for 1 year	Nonfatal myocardial re-infarction	1 in 50	1,119	Coronary Drug Project [62]
CLOFIBRATE 1.8g daily for 1 year	Gallbladder disease	1 in 500	1,103	As above [418]
NIACIN 3.0g daily for 1 year	Arrhythmia	1 in 100	1,119	As above [418]
TOLBUTAMIDE 1.5g daily for 1 year	Cardiovascular death	1 in 70	204	University Group Diabetes Program [63]
PHENFORMIN 100mg daily for 1 year	Cardiovascular death	1 in 60	204	As above [419]
CLOFIBRATE 1.6g daily for 1 year	Total mortality	1 in 900	5,331	WHO clofibrate trial [74]

University Group Diabetes Program trial [63], there was an excess of cardiovascular deaths in patients given tolbutamide and an excess of one in 60 in those given phenformin [419]. In the World Health Organization clofibrate trial [74] there was an increase in total mortality in patients given clofibrate. Oliver has argued that the excess mortality may not be a direct consequence of the action of clofibrate but could result from the lowering of serum or membrane cholesterol even when achieved by dietary means [420].

The large clinical trial is a very powerful tool for detecting adverse drug effects. However, as mentioned earlier with the practolol trial, a large trial may fail to detect an ADR if the methods of recording such a reaction are not adequate.

METHODS OF DETECTING AN ADR DURING A TRIAL

The following data should be collected: the fact and cause of any death; the nature of any illness; all hospital admissions or other contacts; and, where indicated, the results of monitoring biochemical and other tests. Determining the fact and cause of death may require the assistance of a National Central Registry such as the Office of Population Censuses and Surveys in the United Kingdom.

Deaths and causes of death

The progress of subjects withdrawn from a trial must be monitored as dropout may be related to an adverse drug reaction or other major end point. The monitoring of defaulters is especially important. A patient in a trial may appear to have defaulted but instead may have died at home or when on holiday. When repeated attempts to contact a defaulter fail, the National Central Registry for Deaths should be consulted, if accessible, to determine whether or not the subject is still alive and, if dead, the cause of death.

Illness, doctor contacts, hospital admissions

Skegg and Doll [409] have argued that the unexpected and unique oculocutaneous syndrome associated with practolol may have been detected during a large trial if data on doctor contacts had been collected. There may have been an increased number of consultations with eye problems. A particular ADR may not be anticipated or looked for during the course of a trial but if data are regularly collected on episodes of illness, visits to medical practitioners for any reason, and hospital admissions for any cause, then an excess of a particular episode may be noted when an ADR is occurring. The monitoring of disease episodes should be sensitive enough to detect an increased consultation rate for such items as influenza which may be a drug-induced fever, and for depression which may be associated with drug administration.

Biochemical monitoring

Where possible, it is obviously preferable to detect changes in biochemical tests rather than an established ADR. Haematological results, renal function, and hepatic and thyroid function tests are monitored in most trials. If there is a known probability of an adverse reaction then specific biochemical results can be monitored. For example, a long-term trial in the elderly revealed a deterioration in glucose tolerance with diuretic treatment before the occurrence of clinical diabetes [85].

Other investigations

With certain treatments it may be appropriate to monitor other bodily functions; for example, using repeated electrocardiograms, chest radiographs, peak ventilatory flow measurements, or examinations of the optic fundi.

CONCLUSIONS

The methodology for detecting an ADR in a clinical trial must be improved so that we can be confident that drugs monitored in large long-term trials are safe and to be preferred to those not so evaluated. However, some ADRs will be very difficult to detect, especially changes in affect such as depression or anxiety. How will we determine that a new drug produces schizophrenia in one-half of one percent of cases? The large randomised controlled trial with a full documentation of all disease episodes provides the best hope.

17. MEASUREMENT OF THE BENEFIT : RISK COMPARISON, COST-EFFECTIVENESS AND COST-UTILITY.

The benefits of treatment comprise reductions in mortality and morbidity and improvements in quality of life (QOL). The risks should be expressed in the same units as benefits and comprise increases in mortality and morbidity and a deterioration in QOL. Both benefit and risk should be expressed numerically as incidence, for example a 1% decrease in mortality from coronary heart disease/year to be compared with 0.01%/year increase in deaths from liver disease.

Table 17-1 gives a benefit : risk analysis for the treatment of diastolic hypertension. The results are reported for one trial in the elderly, the European Working Party on High Blood Pressure in the Elderly (EWPHE trial [60]) and one in middle aged subjects, the Medical Research Council (MRC) trial [181].

In the EWPHE trial, the benefits were striking and the risks of dry mouth and diarrhoea were mainly due to using methyldopa, a drug that is hardly used now [184]. The benefits outweighed the risks and it is widely agreed that the elderly with high diastolic and systolic pressures should be treated [184]. Why, however, are we treating the middle aged with hypertension? Small benefits of stroke event reduction (1.2/1000/year) and cardiac events (0.3/1000/year) have to be weighed against side effects as high as 11/1000/year for impotence in men. We are probably treating the middle-aged with mild hypertension because most believe that the benefits are underestimated in trials and the risks are exaggerated. The benefits may increase with time, any improvement in the cardiovascular system taking time to take effect, and benefits should certainly accumulate from year to year. Moreover the benefit of active treatment is underestimated because subjects at greatest risk leave the placebo group. Also death and stroke

BENEFITS (per 1000/year)	RISKS (per 1000/year)

SYSTOLIC +
DIASTOLIC HYPERTENSION

Elderly (EWPHE trial)

Fatal events		Fatal events	
Cardiac	-11	None	
Stroke	-6		

Non fatal events		Non fatal events	
Stroke	-11	Diabetes	+9
Severe heart failure	-8	Gout	+4
Severe retinal changes	-4	Dry mouth	+124
		Diarrhoea	+71
		Nausea	[+6]
		Dizziness	[+6]
		Muscle cramps	[+5]
		Lethargy	[+4]

DIASTOLIC ±
SYSTOLIC HYPERTENSION

Middle age (MRC trial)

Fatal events		Fatal events
All	-0.1	None

All events		Non fatal events		
			Diuretic	Propranolol
Strokes	-1.2	Diabetes	+4	0
Cardiac	-0.3	Gout	[+4]	0
		Impotence(men)	+11	+5
		Lethargy	+2	+6
		Dyspnoea	0	+7
		Raynaud's	0	+5

Table 17-1. Benefit and risk analysis for treating mainly diastolic hypertension in the elderly and middle aged. Men and women combined. Treatment was diuretic ± methyldopa in elderly, diuretic or propranolol in middle age. [ ] = estimate.

events could not be more serious and are worth accepting risks to prevent. In addition the clinician may often see patients who are expected to benefit more than the average trial patient. The MRC trial identified men and those with higher pressures as experiencing greater benefits.

The risks may also be exaggerated as they may not accumulate with time. For example if a subject is going to experience an adverse symptom or biochemical side effect this may be experienced in the first year, with a reduction in incidence thereafter. Moreover many symptom complaints may be trivial in comparison with, say, a stroke.

Table 17-2 gives the benefit and risk analysis for patients with systolic hypertension as the main entry criterion. The MRC elderly trial entered patients with systolic hypertension but allowed patients to enrol with a high diastolic pressure [328], whereas the Systolic Hypertension in the Elderly Program (SHEP) trial entered only those with isolated systolic hypertension [182]. In these elderly subjects the benefit appears to outweigh the risks. However, anxiety has been expressed by Davey Smith and Eggar [185] that the benefits only exceed risks at higher levels of systolic pressure. The trial included patients with a systolic pressure of 160 mmHg but possibly only those with a pressure of, say, 180 mmHg have a favourable benefit : risk comparison.

Doubts on the advantages of treatment have not only been expressed in the field of hypertension. Dollery [421] considered that the risks reported in the University Group Diabetes Project (UGDP) trial [63], and the WHO clofibrate trial [74] outweighed the benefits. As discussed in chapter 16, the adverse events in these two trials were severe, therefore the drugs were not effective for their primary purpose and increased overall mortality. Dollery also expressed concern that drugs similar to those discredited in these trials are now being employed without evidence of a satisfactory benefit : risk comparison, and with respect to the MRC trial in middle aged

BENEFITS (per 1000/year)	RISKS (per 1000/year)

**SYSTOLIC ±
DIASTOLIC HYPERTENSION**

Elderly (MRC trial)

Fatal events

    All   -0.8

All events

    Stroke -2.7
    Cardiac -1.9

Fatal events

    None

Non fatal events

	Diuretic	Atenolol
Diabetes	+4	0
Gout	+4	+3
Dyspnoea	+1	+22
Lethargy	+4	+17
Dizziness	+6	+9
Raynaud's	+1	+11
Nausea	+6	+3
Headache	+3	+6
Muscle cramp	+5	+1

**SYSTOLIC HYPERTENSION
(NO DIASTOLIC HYPERTENSION)**

Elderly (SHEP trial)

Fatal events

    Cardiac -2
    Stroke -0.4
    All   -2.7

Fatal events

    None

Non fatal events

    Stroke   -5
    Myocardial infarction -2
    Heart failure   -5
    Coronary artery bypass
     graft   -2

Non fatal events

    [estimated as above]

Table 17-2. Benefit and risk analysis for treating systolic hypertension in the elderly. The MRC treated either with a diuretic or atenolol. The SHEP trial mainly with a diuretic ± atenolol.

subjects he stated *"we have a reduction in stroke and some indication in one sub-group of a reduced mortality from myocardial infarction, while on the other hand there is a clear hint that overall mortality is increased in women and that there is a fairly high incidence of gout, diabetes, nausea, dizziness, impotence and so on. With all that information, it is still very difficult to make a sensible recommendations about treatment policy".*

These questions cannot be easily answered without a full QOL assessment. If QOL has been fully evaluated then survival may be adjusted for QOL to give, for example, QOL adjusted years of survival, or QALYS as they have been termed (see below).

THE COST-EFFECTIVENESS OF TREATMENT

The cost-effectiveness of treatment is the cost of obtaining a certain biological response. At the most simple level, let C = costs; B = economic benefits; and E be the biological response; then:

$$\text{cost-effectiveness} = \frac{C - B}{E}$$

E may be any biological response, for example degree of lowering of blood sugar due to drug treatment or number of years of life gained. Economic analyses have been criticised from many aspects, mainly if the measurement of biological response is not acceptable [422]. For example, epidemiological data may be employed to estimate the extra years of life gained from the treatment of diabetes mellitus. It may be known that subjects with a fasting blood sugar lowered by 2 mmol/l live an extra *x* years. However, to determine the cost-effectiveness of, say, drug treatment, it has to be assumed that the drug will indeed lower the blood sugar by 2 mmol/l and most importantly, that this will result in the extra *x* years of life. This is obviously most unlikely as the effects of the higher blood sugar *in the past* may not be rectified immediately or in the future. This can be allowed for to some extent by assuming the fraction of benefit (FOB) that will occur but data are required to estimate the FOB and to decide when any benefit is likely to be seen.

Johannesson and Jönsson [423] recognised that problems resided in 6 main areas: the epidemiological data (reliability and generalisability), the effectiveness of treatment (the FOB), the outcome measure (eg. life years gained), the costs to be included, the discounting of benefits as costs may be incurred now and benefits to be observed in the future, and the duration of therapy, for example should life long treatment be considered or treatment during a certain period? However, in a randomised controlled trial against placebo the effectiveness is observed, not assumed, the duration of treatment is known and the major difficulty is in determining costs and discounting. Discounting (adjusting) may be necessary as it allows for the fact that we prefer a) to pay later and b) to receive benefits immediately. Costs and benefits may therefore be discounted to allow for this. Nevertheless, Drummond [424] considered that discounting would have *'little quantitative impact on a comparison of two interventions with a similar time profile of costs and consequences.'* However he pointed out that discounting would be necessary if a curative programme was being compared with a preventive one. The former programme would produce benefits immediately and in the latter the benefits would be delayed.

Drummond has written extensively on how and when to determine costs in a controlled trial [424]. He considered that costs should be collected and a cost-effectiveness analysis made when:

i) the trial is well designed and expected to give unbiased and unambiguous answers.
ii) the interventions are of greatly different costs (an analysis is only necessary when the more expensive treatment is more effective).
iii) economic benefits may be expected.
iv) the trial compares a new treatment with the old (there would be cost implications from

v)   the trial is conducted in a typical setting and should provide generalisable results.

vi)   the collection of economic data will not seriously overburden investigators or patients.

Drummond also discussed what costs should be assessed and how, and he also expressed caution in three areas. First the clinical trial results are unlikely to be quantitatively the same in practice owing to following a different protocol and to lower compliance with treatment by the non trial patients. Second the costs observed in the trial will not be experienced elsewhere as costs and resources will vary from place to place. Third the treatment will probably be applied to different categories of patients from those in the trial and the effectiveness of treatment will differ. Nevertheless there is growing pressure on health care budgets and value for money is becoming increasing important. Measuring cost-effectiveness in a trial should often be considered and a health economist consulted.

THE COST-UTILITY OF TREATMENT

A cost-effectiveness analysis may not be adequate under certain circumstances. For example chemotherapy may prolong life in cancer to give a cost per life year gained. However, these extra years may be of poor quality and not worth the money. 'Utilities' are a set of values for health states and the extra years may be adjusted using these utilities to give a quality of life adjusted life year or QALY. The cost-utility of a treatment is therefore equal to $\frac{C - B}{U}$ where C and B are costs and economic benefits as before and U is the life gain adjusted for quality. In chapter 15 we introduced the concept of a Health Status Index or HSI which ranges between 1 (perfect health) and 0 (death). If an extra year of life is multiplied by the HSI then we have a quality of life adjusted year of survival or QALY. Obviously if the patient was perfectly well

Table 17-3. Health states or states of well-being (after Fanshel and Bush [393]).

Health state		Score
S_1	*Total well-being*	1.0
S_2	*Minor dissatisfaction* Very slight but significant deviation from well-being (eg. carries glasses for reading).	0.975
S_3	*Discomfort* Subject has a symptomatic complaint with no significant reduction in efficiency.	0.875
S_4	*Minor disability* Daily activities continue but with a significant reduction in efficiency.	0.8
S_5	*Major disability* Patients show a severe reduction in efficiency of usual functions.	0.75
S_6	*Disabled* Unable to go to work but can get about in the community.	0.625
S_7	*Confined* In an institution.	0.375
S_8	*Bedridden*	0.125
S_9	*Isolated* For example, in intensive care.	0.025
S_{10}	*Comatose*	0
S_{11}	*Dead*	0

after 1 year he would have 1 QALY and if he died immediately, a zero QALY. However, the states between these extremes need to be determined and numerical values placed on these states. Following the suggestion of Fanshel and Bush [393] table 17-3 lists eleven states of well-being and the scores that may be attached to each state. The scores are based on the assumption that a patient is prepared to trade a certain number of years of life in a reduced state of health for a smaller number of years of life in an improved state of health. The scores are somewhat arbitrary and open to discussion. For example, the health state was calculated on the assumption that a person aged 40 would consider a further 40 years of life in a disabled state to be equal to 25 years of life in a state of total well-being.

Estimates then have to be made of any disablement that would prevent a patient's mobility or ability to work, any disability interfering with other aspects of a subject's life, and any discomfort such as that produced by minor symptom side effects. Although the HSI has proved useful in comparing treatments in hypertension [422] and angina [391] QALYS have also been costed and compared for such different treatments as hip replacements and renal dialysis [425]. With the first treatment the major impact is on QOL and with the second treatment, years of life gained. With such analyses it is no surprise that Fletcher has stated "there are problems with current methodology to suggest we should not be using QALYS to determine treatment options" [426]. Acceptance of this view does not imply that QOL cannot be measured and compared within a double blind randomised controlled trial. If such a trial includes an overall measure of QOL, such as a HSI, mortality as an outcome, and different treatments then the trial would generate QALYS which could be employed to decide on the best treatment used in the trial. Unfortunately there is probably no trial as yet that has measured survival, costs and QOL carefully enough to allow this comparison.

Lastly it must be pointed out that QALYS may be calculated assuming that the HSI is constant over a considerable period of time and this may or may not be the case. For example an individual may learn to cope with his or her disability and the QOL may improve. The converse may also occur. To avoid this problem it has been suggested that the Healthy Years Equivalent (HYE) should be calculated [427]. However as Drummond has stated 'More empirical research is needed in order to assess whether HYEs differ from QALYS and, if so, in which situations'.

SUMMARY

The results of a randomised controlled trial may allow us to make a benefit : risk comparison. How such an analysis should be performed is discussed and the methods illustrated for the treatment of hypertension. The expense of treatment is an important issue and if costs are known or collected during the course of a trial, a cost-effectiveness analysis can identify the cost of, say, lowering blood sugar by a given amount or prolonging life by one year. If the years of life gained are of reduced or increased quality, they may be adjusted for QOL to give a quality of life adjusted year of survival (QALY) or a healthy years equivalent (HYE). The cost of 1 QALY or HYE may then be costed in a cost-utility analysis. These evaluations have many problems but may be particularly appropriate when comparing outcomes from different treatments in a controlled trial. Their usage may be expected to increase in the future, especially when costs, QOL and survival are compared in one large trial.

18. EARLY TRIALS ON NEW DRUGS

This chapter considers the ways in which early trials of new drugs differ from trials of established treatments. Greenwood and Todd [428] have defined three phases of early trials: trials to determine safety and early clinical pharmacology (phase I); trials to determine clinical efficacy and further clinical pharmacology (phase II); and trials for the early clinical development of the drug (phase III). A drug regulatory authority may be involved in these early trials.

APPROVAL BY A REGULATORY AUTHORITY FOR EARLY TRIALS

Until the early 1980s the usual form of authorization in the United Kingdom was via a clinical trial certificate. The manufacturer of the new drug applied to the then Medicines Division of the Department of Health and Social Security (DHSS) giving the chemistry, pharmacology, and the details of animal experiments with the drug [429]. The DHSS division reported to the Committee on Safety of Medicines (CSM) with advice on whether or not a clinical trial certificate should be issued.

Outside the United Kingdom, it was often only necessary to inform the regulatory authority of the intention to perform trials on a new substance or new clinical entity (NCE). No certificate was issued, but the regulatory agency could object. In the United States a notice of Claimed Investigational Exemption for a New Drug was filed with the Food and Drug Administration (FDA). Simon and Jones summarised the countries requiring only notification, and these included most European countries. They also listed the countries requiring a formal detailed submission and approval by a regulatory authority, for example, Australia, Canada, India, Israel, and South Africa [429].

The manufacturer of a new drug is very concerned with shortening the development time and the regulations in the United Kingdom led to long delays (sometimes over eight months). The number of clinical trial certificates issued fell from over 170 per year in 1972-1974 to 87 per year in 1980 [430]; early trials were conducted in other countries. In 1981 a new scheme was introduced where exemption from the need to obtain a clinical trial certificate may be granted when the licensing authority receives the following:

1. certified summaries of the basic data
2. a copy of the trial protocol
3. confirmation that a medical adviser to the company, working in the United Kingdom, is satisfied that the trial is reasonable

If the licensing authority objects within 35 days, the pharmaceutical company could still apply for a clinical trial certificate as before [430]. The relaxation of regulations would appear reasonable on two counts: first, exemption has always been possible for doctors and dentists conducting trials on their own initiative and second, phase I trials appear to be very safe.

PHASE I TRIALS

Phase I trials tend to be open studies of a new clinical entity and, when not a randomised controlled trial, fall outside the scope of this book. However, a randomised controlled trial is often performed at this stage. For example, a tranquillizer or antihypertensive drug may be expected to have sedative properties and the dose may be increased stepwise in a trial to determine whether the therapeutic effect occurs at a lower dose than the side effect of sedation. In such a study, with both subjective and objective assessments, a randomised controlled trial is appropriate with placebo control. Such a trial differs from studies on established drugs as all adverse effects will be unknown and the trial must be carefully supervised, conducted in a hospital or clinical laboratory, and usually accompanied by haemodynamic and biochemical monitoring. The laboratory must be equipped with all those items that would be required for emergency resuscitation and the staff trained to use this equipment. Written consent is, of course, essential.

Most phase I studies are in healthy volunteers. An obvious exception would be a study of a new toxic anti-cancer agent that would have to be tested in patients. Healthy controls show less physiological variation than patients and may be better able to co-operate in complex experiments [431]. Volunteers are often the staff of a pharmaceutical company or a student and they participate, partly at least, to receive financial rewards to compensate for any inconvenience they may have suffered.

In the early 1980s two volunteers died and later a Research Organisation closed because of an outbreak of hepatitis [431]. However, Orme and colleagues performed a survey in 1986-87 and found that of 8163 healthy volunteers only 3 had a potentially life threatening adverse event and there were no lasting sequelae [432].

Guidelines have now been published by several bodies [433-436]. In general they address the following issues:

i)   A risk greater than minimal is not acceptable
ii) Informed written consent must be obtained. Any known or suspected risks should be made clear
iii) Financial inducements should not be excessive and initial recruitment should be by announcement or notice and not targeted at an individual
iv) Ethics Committee approval is required
v)   The sponsor must agree to pay compensation if the volunteer suffers any deterioration in well-being as a result of the research [437]
vi) The volunteer should not participate in more than a few studies and clinical problems

that would make the subject ineligible should be sought (with permission) from the volunteer's primary care physician [438].

Phase I trials are very important in determining tolerability, the dose response curve and the concentration-activity relationships. The trial may detect adverse effects that occur almost universally and examples include sedation, nausea and adverse cardiac effects [439]. A dose response relationship may be apparent for these effects in Phase I trials but this problem is usually explored in patients. The concentration-activity relationships are mainly concerned with determining the degree of absorption, the variability in absorption, route of elimination, metabolite formation and activity, and bioavailability.

Phase I studies determine whether the new drug has a pharmacological action that may be useful in treatment, and phase II trials examine whether this action proves a benefit in patients with disease [440].

PHASE II TRIALS

Phase I studies, having provided data on the safety and clinical pharmacology of the new drug, are extended in phase II in order to determine clinical efficacy in patients and the doses to be employed. These studies may well be randomised controlled trials, with double-blinding and conducted on outpatients. Careful monitoring for biochemical and other adverse effects will be necessary. In addition it is important to determine the 'minimal effective dose'. The desirability, from an ethical point of view, of determining the 'maximum tolerated dose' remains an open question.

PHASE III TRIALS

Greenwood and Todd [428] have considered certain objectives for phase III trials: definition of those patients who would benefit from the use of the drug; comparison of the new drug with existing drugs; detection of less common adverse effects; determination of any tolerance to the drug's effect, detection of interactions with other drugs, tobacco, and alcohol; the use of the drug in geriatric and paediatric patients; and further studies on the mode of action. Trials in phase III may differ little from the standard randomised controlled trial.

REGULATIONS GOVERNING WHETHER A NEW DRUG CAN BE GENERALLY RELEASED

Before the drug regulatory authorities of different countries will authorise the release of a new drug they must be satisfied about the efficacy, safety, and quality control of the product. Norway also requires a medical need for the new drug to be demonstrated. The activities of these authorities have been reviewed by Lumbroso [441].

Stringent clinical trials are now required by all authorities, but in the past countries have varied widely in their requirements. Regulatory authorities were mainly established after therapeutic disasters; for example, in 1937 the Food and Drug Administration (FDA) was created in the United States of America following deaths from a sulphanilamide elixir containing ethylene glycol. In France the regulations were strengthened in 1952 following deaths from a preparation of diethyl tin diiodide. The thalidomide disaster in 1959-1960 led to regulatory authorities being established in many countries. The Committee on Safety of Drugs in the United Kingdom was established in 1964 and involved a voluntary system that became law between 1968 and 1971. Interestingly, Lumbroso points out that in Western Germany (where thalidomide was developed) regulations were first imposed by the EEC in 1972.

The regulatory authorities still vary in the amount and type of information required. One of the most cautious authorities is the FDA, whose deliberations may delay the introduction of new drugs by more than three years. The FDA also requires a copy of all record forms completed during the course of clinical trials. The strict regulations are designed to prevent the introduction of a potentially dangerous drug but it is admitted that, in the process, the public may be deprived of a beneficial drug for a period of time.

Some countries also require research to be replicated in their country. Lumbroso [441] feared that this requirement could be misused for a commercial protectionist purpose but pointed out that the replication of studies can bring to light new therapeutic indications and clarify the action of new drugs. However, this additional information may better be sought from different and specially designed studies. The difficulty of devising an internationally acceptable standard trial protocol for new drugs arises from differences in attitudes and legal standards between countries rather than disagreements on scientific merit [442]. Whether a drug will be marketed in a particular country depends not only on the absolute safety in, say, deaths per 10,000 treatments, but also to the country's attitude to such deaths. Where death can follow an infectious disease, an inexpensive antibiotic with some risk is better than a treatment that is safer but too expensive to be provided. However, the promotional activity of pharmaceutical companies may determine which drug is prescribed rather than considerations of cost effectiveness [443].

POSTMARKETING TRIALS

Clinical trials on a new drug do not cease with registration for sale. Very important trials may be started after this event in order to examine the long-term efficacy of treatment. Also, prior to registration, it is difficult or impossible to observe a sufficient number of patients for a long enough period to detect rare adverse reactions. It is hoped to overcome this difficulty by monitoring adverse effects in postmarketing surveillance schemes. However, the large randomised controlled trial provides the best opportunity of detecting adverse drug effects as in the trials of clofibrate and oral antidiabetic drugs discussed in chapter 16. Important randomised controlled trials after marketing include further trials of efficacy in comparison with other drugs, more trials to detect interactions, further trials in certain groups such as the young and old, and more trials to determine the optimum dose and dose frequency. Surprisingly, in the past little has been known about these aspects at the time of registration. Most important is the trial that goes beyond the surrogate end-point, for example lowering of blood pressure, blood cholesterol or blood sugar to see whether the promising pharmacological action leads to a true reduction in mortality and morbidity. Without such a trial doubts may be expressed about the value of such treatment.

Postmarketing trials have also been termed phase IV studies. One particular variety is currently promotional and has been devised to familiarise clinicians with the use of the drug. Such studies do not usually employ a randomised control group and should only be supported when the drug represents an important new treatment. However, they are usually employed to introduce a further sedative, antiinflammatory, or anti-hypertensive drug and the purpose of the trial is to sell the product. The trials are often carried out in general practice and inducements have been provided to persuade a large number of doctors to use the new drug. At the end of such a study the patient may continue to receive the drug at the patient's or government's expense and the pharmaceutical company may recover more than the cost of the trial. An article in the *Sunday Times* of January 29, 1978 summed up the situation.

Patients put at risk as doctors aid firms in sales drive

The Sunday Times interviewed 39 GP's. Four admitted that they do not tell the patient or ask permission when they are testing a new drug. Twenty-seven said taking part in the trial had influenced their choice of drug, and some said they would never have chosen the drug for the patient if they had not been asked to test it.

This article questioned the ethical position of certain doctors involved in these trials and then criticised the trials for failing to collect important data on adverse events and for paying the doctors to take part. The advantages and disadvantages of promotional trials deserve further attention and their usefulness may depend on the drug being investigated.

Lionel and Herxheimer [444] have stressed that a *"good proportion of the drugs available are of little importance in terms of essential health care and are marketed mainly because they can be sold and not because they benefit the health of the population."* We should concentrate the limited and valuable resources for randomised controlled trials on fewer drugs.

The clinical investigator should be less willing to investigate a new drug when it is closely similar to many that are already available. These compounds have been termed 'me-too' drugs and resources may be better employed in examining more original drugs. However, it must be admitted that some 'me-too' drugs prove to have a unique place and to represent a real advance.

LIMITATIONS OF TRIALS IN NEW DRUGS

Small trials during the early phases of drug development may fail to detect severe adverse reactions and new drugs may be released and subsequently withdrawn from the market. This happened with practolol, even though the drug had been used in one large randomised controlled trial involving over 3,000 patients [414].

CONCLUSIONS

Randomised controlled trials are necessary for the evaluation of new drugs and this chapter considered the type of trial appropriate for each phase in the investigation of a new compound. The contribution of regulatory authorities was discussed as they have to approve new drugs for use in trials or for general release.

Clinical investigators should be more discriminating in the type of trial in which they become involved. They should avoid promotional trials of 'me-too' drugs which originate from the marketing departments of pharmaceutical companies and should take part in trials of potentially important new drugs or trials of established drugs where an attempt is being made to answer an important question.

19. FAILURE TO ACCEPT THE RESULTS OF RANDOMISED CONTROLLED TRIALS

The result of a trial may not be accepted for several reasons: the result may be at variance with preconceived ideas; an unusual group of patients may have been recruited; the treatment groups may not be identical in important respects; too few patients may have been recruited and the power of the trial may be too low; the results of the trial may not have been interpreted correctly; the trial result may not be consistent across different strata of patients; the trial may provide a result that conflicts with the results of other trials; the treatment may be difficult to administer or have too many adverse effects; the trial design may be faulty; and finally, the trial may originate from a group with a vested interest in demonstrating the observed result (for example, a pharmaceutical company). Before discussing each of these reasons we shall illustrate them by describing trials whose results were not completely accepted (at least at first) and also series of trials on related drugs, the collective results of which have been difficult to interpret. After discussing these trials, we shall return to the reasons for rejecting the results of a randomised controlled trial.

THE ANTURANE REINFARCTION TRIAL

The Anturane Reinfarction Trial was a randomised, double-blind, multicentre trial comparing sulphinpyrazone (Anturane, 200 mg four times a day) with a placebo for the secondary prevention of myocardial infarction. The trial started in September 1975 and an interim analysis in July 1977 revealed a statistically significant (P ≤ 0.02) reduction in cardiac deaths after an average follow-up period of 8.4 months. The results were published in February 1978 [65] and

Table 19-1. Results of the Anturane (sulphinpyrazone) trial, made available in two reports [65,445].

	Placebo group	Sulphinpyrazone group	% reduction	P
First report				
Number patients	742	733	-	
Total mortality (n)	60	40	33	Not specified
All cardiac deaths	?	?	?	-
Cardiac deaths analysed	44	24	49	0.02
Sudden deaths analysed	29	13	57	0.02
Second report				
Number patients	783	775		
Total mortality (n)	85	64	25	Not specified
All cardiac deaths	78	59	24	Not specified
Cardiac deaths analysed	62	43	32	0.06
Sudden deaths analysed	37	22	43	0.04
Sudden deaths (2-7 months)	24	6	74	0.0003
Randomised but ineligible	33	38	-	-
Withdrew from study	220	195	-	-

are given under the heading of 'First Report' in table 19-1. Recruitment to the trial was stopped at the time of this report but the investigators disclosed the short-term results to the subjects in the trial and sought their individual consent to continue. All but seven patients agreed to continue. A second report was published in January 1980 [445] and the results are also given in table 19-1.

A 49 percent reduction in cardiac deaths was reported in the first article and a 32 percent decrease in the second. Unexpectedly, the reduction in deaths was not due to the postulated decrease in further episodes of myocardial infarction but to a reduction in sudden deaths, which were possibly related to arrhythmias. The trial has been criticised and the adverse comments concern the definitions employed in the trial, the organisation of the trial, and the manner in which the trial was published.

Definitions employed in the trial

Ineligible patients
Ineligible patients were those who were randomised into the trial but were excluded from analysis by the policy committee as the patients did not meet the criteria of the investigation protocol. It appears that some patients in this group were excluded after they had died and therefore the definition of an ineligible patient may crucially affect the results.

Nonanalysable deaths
Nonanalysable deaths included those occurring either within the first seven days of starting treatment or more than seven days after stopping treatment. Nonanalysable deaths therefore included all deaths in patients who dropped out and deaths among patients who did not comply with instructions to take their medication. Nonanalysable deaths also included those *"attributed directly to surgery in which no association could be established with a nonfatal event while the patient was on study treatment"* [445]. The analysis was on the on-randomised treatment or per-protocol principle (chapter 14). An intention-to-treat analysis was not performed.

Sudden death
A sudden death was one that was either not observed or one that occurred within 60 minutes of the onset of symptoms.

Organisation of the trial

The organisation of the trial included the following features:

Coordinating centre
The coordinating centre was situated at the Ciba-Geigy Corporation and its Operations Committee was responsible for the execution of trial procedures and the reporting of data to a policy committee. The data were verified by independent university departments of epidemiology and the trial procedures were similarly audited.

Policy, Audit, and Electrocardiographic Committees
These committees were independent of the pharmaceutical company.

Financing of the trial
The Ciba-Geigy Corporation, the manufacturers of Anturane, financed the trial.

Manner in which the trial was reported

The trial was published as two results from analyses at different stages. Therefore multiple looks at the data were made. Table 19-1 provides the results of the trial in the two reports. There was a fair agreement between the effect of Anturane on sudden deaths; the reduction was 57 percent and 43 percent in the first and second reports, respectively. In the second article it was stated that this reduction mainly occurred between the second and seventh months of treatment (74 percent reduction in sudden deaths).

Comments that have been published concerning the trial

Four important reviews are summarised in the following sections.

Editorial in the Review Epidemiologie et Santé Publique
Armitage wrote this editorial in 1979 [446]. First, he considered the problem of repeated looks (chapter 6 and 14) and commented, *"The ART made no explicit allowance for such repeated testing, but it seems likely that for the particular design used, whereby the first examination on the data occurred after a substantial proportion of the total patient intake, the adjustment needed would not be great."* This statement was made before publication of the second report when, presumably, further looks were undertaken.

When discussing the selection of some end points as either analysable or nonanalysable deaths, Armitage stated, *"The investigators have chosen to discard the safe 'pragmatic' approach in favour of an 'explanatory' approach which may be more sensitive to the presence of real effects but may also suffer from bias."*

Editorial in the New England Journal of Medicine
In an editorial in January 1980 Braunwald commented [447], *"It would certainly be desirable to repeat the sulfinpyrazone study to confirm its results..."* but *"... despite the great desirability of learning more about this drug, the information available suggests that sulfinpyrazone should be approved for use after infarction and made available to the American public at the earliest possible time."*
The Food and Drug Administration considered whether or not to license the drug for use in the secondary prevention of myocardial infarction and did not agree [448].

Article in Science entitled "FDA says No to Anturane"
Kolata [448] reported that the U.S. Food and Drug Administration refused to approve sulphinpyrazone for the secondary prevention of myocardial infarction in April 1980 on the grounds that the case for this drug was not persuasive. She quoted some very important comments. Paul Meier made a remark concerning nonanalysable deaths: *"The idea of nonanalysable deaths is an innovation in the analysis of clinical trials that we can do without."* Meier also commented that the exclusion of ineligible patients magnified the differences between the groups in the trial. In defence, Sol Sherry (chairman of the trial's policy committee) pointed out that bias was prevented by the double-blind nature of the trial and that the exclusion of certain patients was done prospectively and not retrospectively after seeing the data.
Kolata quoted an epidemiologist as saying that the report of the study *"was orchestrated [by Ciba-Geigy] for presentation in the scientific and public arena so as to create an impression that there was an unequivocable clear-cut, dramatic result. What happened was almost a con job."* Meier was also reported as saying, *"It was an interesting but not a convincingly positive result. It was made into a break-through by PR [public relations]."*
Most importantly, Kolata reports that Robert Temple, head of the Food and Drug

Administration's cardiorenal division, audited the study and found the following:

i)   The classification of many sudden deaths was incorrect.   A   reclassification   removed   the deficiency   of   sudden   deaths   in   the sulphinpyrazone group. The definition of sudden death *"wasn't very well thought out. It turned out to be more crucial than anyone would have anticipated."*
It has subsequently been suggested by a member of the trial's policy committee that Temple was biased in his reclassification.

ii)   The results were affected when nonanalysable deaths were also included in the analysis. A 32 percent reduction in analysable cardiac deaths with the use of sulphinpyrazone was not altered to a great extent and was lowered to a 24 percent reduction by including nonanalysable deaths.

iii)   The results were altered when ineligible patients were included. Some patients were ruled ineligible after they had died. Temple considered *"everyone in this business knows [such exclusions of dead patients] just are not done"* [448].

Review in the British Medical Journal
Mitchell pointed out that the numerous exclusion criteria resulted in a highly selected trial population and that the results may not be applicable to other groups of patients with myocardial infarction [449]. He was also worried about the analysis of the trial and considered that the only acceptable analysis of outcome is one based on the intention-to-treat principal (see chapter 14). Mitchell concluded, *"For the present, my verdict on the claim that the report of the ART has altered the state of the art must be 'not proven'."*

Conclusions on the Anturane Reinfarction Trial

The basic design of the trial was sound and although the data should have been analysed on both the intention-to-treat and per-protocol basis, I have not seen evidence that an erroneous conclusion was reached. A re-analysis by an independent committee found a non significant trend suggestive of benefit [450]. The original report was of a 51% reduction in total cardiac death and the re-analysis on the on-randomised treatment basis revealed a 45% reduction (P=0.08). The re-analysis on the intention-to-treat basis revealed a 33% reduction in total cardiac mortality (P=0.12). Moreover the Anturane Reinfarction Italian Study (ARIS) [451], reported a 56% reduction in reinfarction with sulphinpyrazone but no reduction in sudden death. In this trial the corresponding intention-to-treat analysis for *all* events (including 1 cancer death and 1 death from peritonitis) suggested a 40% reduction in total non fatal and fatal events.

MULTICENTRE TRIAL OF STREPTOKINASE

The European Co-operative Study Group for Streptokinase Treatment conducted a multicentre trial of 24 hour treatment with streptokinase in patients suffering from acute myocardial infarction [452]. The results were published in 1979. Entry to the trial had to be within 12 hours of the onset of chest pain and only 13.5 percent of 2,338 patients with suspected infarction could be entered into the trial. After six months, 48 control patients and 24 treated with streptokinase had died. Table 19-2 gives the deaths according to the time after treatment and reveals the largest benefit from 21 days onwards.

Table 19-2. Results of the European Co-operative Study Group for Streptokinase Treatment trial according to the time after treatment [452].

Time after treatment	Deaths in the control group	Deaths in the streptokinase group
0-21 days	28	18
21-183 days	20	6

Critical leading article

The results were criticised in a leading article in the *British Medical Journal* [453] for the following reasons:

i)   The infusion of a lytic agent for the first 24 hours could not influence late mortality, and this was the mortality most affected.

ii)   Random allocation failed to balance out all the highly relevant risk factors. In the streptokinase group, fewer patients had suffered a myocardial infarction in the past and fewer developed dysrhythmias in the coronary care unit. (The latter may possibly have been due to streptokinase treatment).

iii) There was a low randomisation rate into the trial and many patients were excluded.

iv)   The leading article went on to consider the merits of the trial and to discuss the possibility that the results were real. However, the trial results were criticised for not fulfilling the author's expectations and producing results that were difficult to implement. To quote, *"... the clinical and laboratory complexities inherent in any effective and well-controlled lytic regimen will limit the practical impact of the study on doctors ..."* [453].

Comment

The criticisms of the multicentre trial do not appear too serious. Low randomisation rates are a common and usually unavoidable problem when early treatment is required, and the assumption that streptokinase could not affect late mortality cannot be supported as acute treatment may limit the size of the infarct. Random allocation often results in groups that are unequal in a small number of respects, but an adjustment can be made during analysis. Finally, although lytic treatment is a difficult procedure the drug has been infused directly into the coronary arteries - a much more onerous task.

The use of streptokinase in acute myocardial infarction - an overview

It is interesting that the 'rejection' of the results of this trial was so misplaced. Fletcher *et al* [454] first reported the use of streptokinase in acute myocardial infarction (AMI) in 1958. After 31 trials including 41,000 patients it appears that streptokinase reduces mortality by 24% [455]. The ISIS 2 trial included 17,000 cases treated within 24 hours (median 5 hours) and reported a 25% reduction in 5 week vascular mortality [77]. The efficacy of streptokinase in this situation is no longer disputed. The issue has been resolved by meta-analysis and a megatrial.

TRIAL OF ASPIRIN IN THE SECONDARY PREVENTION OF MYOCARDIAL INFARCTION

In 1980, Elwood [456] summarised the results of six randomised controlled trials of aspirin against placebo in the secondary prevention of myocardial infarction. Table 19-3 is derived from his work. Five of the trials demonstrated a reduction in total mortality of between 15 and 30

Table 19-3. Summary of total mortality in six randomised controlled trials of aspirin in the secondary prevention of myocardial infarction (derived from Elwood [456]).

Trial	Year	Refer-ence	Total no in trial	% of deaths on aspirin	% of deaths on placebo	Effect of aspirin
Elwood and associates	1974	457	1,239	7.6	9.8	-22%
Coronary drug project	1976	458	1,529	5.8	8.3	-30%
Uberla	1978	459	626	8.5	10.3	-18%
Elwood and Sweetnam	1979	460	1,682	12.3	14.8	-17%
PARIS I	1980	461	1,216	10.9	12.8	-15%
AMIS	1980	462	4,524	10.8	9.7	+11%

percent whereas the sixth and largest found an increase in total mortality of 11 percent [462]. However, this large trial (the Aspirin Myocardial Infarction Study) did find a decrease in nonfatal infarctions of 22 percent.

The average reduction in mortality with aspirin was 15 percent (8 percent if the result is adjusted according to the numbers in the trials). In 1980 aspirin appeared to have a small effect that is difficult to detect in randomised controlled trials. However this proved to be an incorrect view.

These early trials employed aspirin up to 5 years after the myocardial infarction [461,462]. The megatrial ISIS 2, not only tested the effect of streptokinase within 24 hours of an AMI but also the effect of aspirin using a factorial design (chapter 5). The acute use of aspirin reduced 5 week vascular mortality by 23% and in the streptokinase plus aspirin group mortality was reduced by 42% [77].

Some controversy still remains over the use of long-term aspirin. However data from overviews suggest that in secondary prevention there is a reduction in vascular mortality of 11% and non fatal infarction is reduced by 31% [455,463].

TRIALS OF DIPYRIDAMOLE POST MYOCARDIAL INFARCTION

In contrast to the acceptance of the use of streptokinase and aspirin after a myocardial infarct (MI) and some support for the use of sulphinpyrazone, the administration of dipyridamole has not been accepted. To quote from the Drug and Therapeutics bulletin [464] *"Such treatment is unjustified in patients who have had a stroke, transient ischaemic attack, or myocardial infarction; promotion encouraging such use is misleading and should not be allowed under the Medicine Act"*. The reasons for such a statement are as follows:

i)   Only one trial compared dipyridamole alone with placebo in AMI. The trial was hopelessly lacking in power, only included 120 patients and failed to reveal any benefit from treatment [465].

ii) The larger trials have consistently employed dipyridamole *in combination with* aspirin. For example, the PARIS I trial (Persantine Aspirin Reinfarction Study [461]) compared 810 patients on dipyridamole plus aspirin, with 810 on aspirin alone and 406 on placebo. The results were almost identical in the two actively treated groups. The PARIS II trial compared 1563 patients on dipyridamole plus aspirin and 1565 on placebo [466]. Coronary deaths were reduced by only 6% although coronary events were reduced by 24%. These effects could easily be attributed to the aspirin treatment [467].

iii) The drug has side effects including headache, nausea and abdominal pain [464,467].

iv) Misleading promotion for the proprietary brand of dipyridamole, Persantine. The slogan was employed *"When thrombosis threatens... Persantine 100 Antithrombotic offers a real opportunity for prevention."* [464].

v) Variable absorption kinetics [467].

vi) Cost [464,467].

Fitzgerald therefore agreed with the Drugs and Therapeutics bulletin and stated *"dipyridamole contributes little if anything to the antithrombotic action of aspirin"* [467].

TRIALS OF ANTICOAGULANTS IN THE SECONDARY PREVENTION OF MYOCARDIAL INFARCTION

In 1970 an international anticoagulant review group combined the results of nine trials of long-term anticoagulant administration after myocardial infarction [76]. The pooled results are given in table 19-4 for males and females separately. Total mortality was 20 percent lower in men given anticoagulants (P <0.01) but only 8 percent lower in women. It is easy to understand why anticoagulant therapy was abandoned for women in most countries, but why was this treatment rejected for men? Anticoagulant therapy for myocardial infarction continued in the Netherlands and was used for both men and women, even the elderly [468]. In most countries the gains from therapy were not thought to be worth the difficulty of administering anticoagulants. To quote Mitchell [449], *"... even if the claims (for anticoagulants) were valid the apparent benefit was too small to justify the hassle of conventional anticoagulant regimens."*

A 20 percent reduction in mortality was not considered worth continuous anticoagulant therapy. This treatment involves repeated estimations of blood coagulability, and bleeding may occur as an adverse effect of too much treatment. The occasional patient may therefore die as a result of treatment and it may be unacceptable to the prescribing clinician to cause one death through treatment even though he may witness several deaths that may (but often may not) have been prevented by treatment.

Of the 16 members of the International Anticoagulant Review Group [76], 14 considered *"... the findings warranted a conclusion that anticoagulant therapy probably prolonged survival at least over two years but that benefit was largely restricted to patients with a history of anterior or previous infarction."*

The remaining two members *"were not convinced that long-term therapy prolonged survival."*

Since 1970, three meta-analyses have been performed, in 1977 [469], 1978 [470] and 1981 [471]. They concluded that anticoagulants reduce mortality. However we still had the Norwegian Warfarin Reinfarction Study (WARIS) trial reporting in 1990 and demonstrating a reduction in mortality of 24% and a reduction in reinfarction of 43% [472].

In the past we can only conclude that the clinicians did not support the burden of anticoagulant therapy. However anticoagulant treatment is routine in many conditions such as artificial heart valves and atrial fibrillation and there must have been doubt about the benefit : risk comparisons reported in the post MI trials, especially the risk of bleeding. To reduce the risk of bleeding, aspirin therapy was not allowed in the WARIS trial. In 1996 it is possible that clinicians consider that aspirin

Table 19-4. Summary of the results of nine controlled trials of long-term anticoagulant therapy in the secondary prevention of myocardial infarction, prepared by an International Anticoagulant Review Group [76].

	Control group			Anticoagulant group			Reduction in total deaths
	PYE	Deaths	Deaths per 100 PYE	PYE	Deaths	Deaths per 100 PYE	
Males	3591	340	9.5	3947	300	7.6	20%
Females	384	43	11.2	476	49	10.2	8%

PYE = person years of exposure

therapy will confer the same benefits. However, the jury has decided on the anticoagulant issue. These drugs should have been employed post MI but they were not.

TRIALS OF BETA-ADRENOCEPTOR BLOCKING DRUGS IN THE SECONDARY PREVENTION OF MYOCARDIAL INFARCTION

Vedin [473] gave an address at an international conference in 1980 and considered the results of five trials of beta-adrenoceptor blocking agents in the secondary prevention of myocardial infarction. Table 19-5 reproduces the results of Vedin's review in which he successively pooled the level of significance for these trials and concluded that beta-adrenoceptor blocking drugs (or at least alprenolol and practolol) were effective in reducing total mortality after a myocardial infarct. Moreover, he suggested that these data were so conclusive that it may be unethical to allow any further placebo-controlled trials on this subject. "nearly 20,000 patients round the world are either already enroled in or will be enroled in prospective secondary prevention trials with beta-blockers. This massive program-costing an estimated 30 million dollars a year is unlikely to benefit either patients or science..."
Reviewing the same data in 1981, Hampton [478] stated, "In 1965, Snow described a clinical trial of propranolol in patients with acute myocardial

Table 19-5. Summary of the results of 5 trials using beta-adrenoceptor blocking drugs in the secondary prevention of myocardial infarction. Successively pooled P values according to Vedin [473].

| Authors | Ref | Drug | Total numbers | | Total Deaths | | Successively pooled P values | |
			Active	Control	Active	Control	Per Compound	Pooled
Reynolds & Whitlock	474	Alprenolol	38	39	3	3		
Ahlmark & Saetre	475	Alprenolol	69	93	5	11	0.41	
Wilhelmsson & others	476	Alprenolol	114	116	7	14	0.10	
Andersen & others <65 yrs	477	Alprenolol	140	142	13	29	0.002	
Multicentre International	414	Practolol	1520	1533	54	83	0.013	0.005

infarction; he found a considerable reduction in mortality ... This was the first post-infarction trial of a beta-blocker, and none of the many subsequent trials have demonstrated such marked benefit." Baber and Lewis reviewed the trials on beta-adrenoceptor blocking drugs and published the 90 percent confidence limits [126]. Of 18 trials, eight had confidence limits encompassing a 50 percent increase in mortality, and 14 a decrease in mortality of 50 percent.

From large trials we now know that beta-blockers should be prescribed post MI [201,218, 479]. Most upsetting, however, was the fact that these benefits were not universally translated into clinical gains. Some cardiologists assumed that the same benefits would accrue for a different class of drugs, the calcium channel blockers. An overview by Held and colleagues [480] and Furberg *et al* [481] suggest that this is not the case.

THE UNIVERSITY GROUP DIABETES PROGRAM TRIAL

The University Group Diabetes Program study was a trial of treatment in newly diagnosed diabetic patients who did not require insulin and had a good prognosis for a five-year survival [63]. Patients were randomly allocated to placebo (PLBO), the oral hypoglycaemic drug tolbutamide (TOLB), a variable dose of insulin (IVAR), or a standard amount of insulin (STD). All patients were given dietary advice. An additional group was randomised to receive phenformin but they were not recruited at the start of the trial and cannot be expected to be identical to the other groups. Table 19-6 gives the results of the trial excluding those in the phenformin group [419].

Table 19-6. Results of the University Diabetes Program (UGDP) Trial.

	Tablet		Insulin	
	Placebo	Tolbutamide	Standard dose	Variable dose
Number	205	204	210	204
Deaths				
Cardiovascular	10	26*	13	12
%	(5)	(13)	(6)	(6)
Non Cardiovascular	11	4	7	6
%	(5)	(2)	(3)	(3)
All causes	21	30	20	18
%	(10)	(15)	(10)	(9)

*$P = 0.005$ compared with the other groups.

There was a statistically significant excess of cardiovascular deaths in the group treated with tolbutamide. Total deaths were also increased in this group but the excess did not reach statistical significance. Not surprisingly since tolbutamide is a popular treatment for diabetes, this trial has been criticised.

Criticisms of the University Group Diabetes Program trial

The conclusion that tolbutamide therapy was associated with an excess of cardiovascular deaths has been criticised on the grounds of inadequate data collection, admission of ineligible patients, administration of a fixed dose of tolbutamide, failure to detect a statistical increase in total mortality, inequality of the groups after randomisation, an abnormal outcome in the placebo group, excess deaths not being observed in every clinic, and patients being transferred from one group to another.

Inadequate data collection
No data were collected on important risk factors such as smoking or the duration of diabetes prior to entering the trial [482,483]. There was no way of knowing whether the groups were comparable for these important factors.

Defence. Randomisation and the numbers in the trial make biologically important differences between the groups unlikely.

Admission of ineligible patients
Certain patients were admitted to the study who should have been excluded on the basis of a poor prognosis for a five-year survival. These patients may have been unevenly distributed between the groups. On the other hand, in many cases the diabetes was very mild or its presence questionable. These patients should only have been eligible for dietary advice and should not have entered the trial. Sixty-nine patients were admitted without meeting a glucose tolerance test criterion [482,483].

Defence. Again, randomisation and the numbers in the trial would make an important and unequal distribution between the groups unlikely. It may be agreed that patients with borderline diabetes should only receive a diet, but that statement can only be made with confidence now that we know the result of the trial. Also, the dilution of the treatment groups with patients who do not have classical diabetes should not bias the results. However, ineligible patients, as defined by exclusion criteria, should not have been included.

Administration of a fixed dose of tolbutamide
A fixed dose of tolbutamide was given to one group of subjects without regard to their individual needs [483].

Defence. The result of the trial for this group is that of a fixed dose. A variable dose may produce a different result but this remains to be proved.

No definite increase in total mortality
There was no statistically significant increase in total mortality in the tolbutamide group and noncardiovascular deaths were reduced in this group.

Defence. Total mortality tended to be higher in the tolbutamide group.

Inequality of the groups after randomisation
The tolbutamide group had more patients with a high serum cholesterol or major electrocardiographic abnormalities, more males, more obese patients, and more with a history of angina [482]. Similar remarks were made about the group randomised to phenformin [484].

Defence. The tolbutamide group included fewer hypertensive patients. Moreover, an adjustment for baseline differences did not materially affect the results [485].

Abnormal outcome in the placebo group
There were no deaths from myocardial infarction in the placebo group [482].

Defence. This could have been a chance occurrence because the mortality from myocardial infarction was very low with this treatment.

The excess deaths were not observed in every clinic
Most of the excess deaths in the group given tolbutamide occurred in only three of the 12 clinics. Schor remarked, *"It would appear to any reasonable statistician that for some reason or other the randomisation procedure broke down in these three clinics over some period of time ..."* [482].

Defence. No evidence has been provided that randomisation broke down. Treatment was assigned by the coordinating centre and not the treating clinic.

Transfer of patients from one group to another
Some patients transferred from one group to another and analyses were only performed on the intention-to-treat principle and not by the per-protocol method.

Defence. There were very few transfers, both kinds of analysis should have been presented. The analysis performed was the least biased.

Conclusions

The University Group Diabetes Program (UGDP) trial was well designed, and the analysis has not been seriously faulted. When a trial shows a possibly beneficial result of treatment it can be repeated; however, it would not be ethical to repeat a trial to confirm a suspected adverse effect. The results must therefore be accepted at this stage, but as patients are still being given hypoglycaemic drugs, it is hoped that observational studies will clarify any adverse consequences of such treatment and indicate which drugs, if any, may be safely prescribed. It may then be possible to arrange a randomised controlled trial on these compounds. A trial could also be restricted to those patients who do not wish to take insulin and do not diet effectively. These subjects could ethically be randomised to tolbutamide or placebo provided they had symptoms from hyperglycaemia that could benefit from treatment and were also informed of the result of the University Group Diabetes

Program trial prior to giving informed consent. This trial would be of possible benefit to the patient and could be justified.

The Committee for the Assessment of Biometric Aspects of Controlled Trials of Hypoglycaemic Agents [486] concluded, *"We consider that in the light of the UGDP findings it remains with the proponents of the oral hypoglycaemics to conduct scientifically adequate studies to justify the continued use of such agents."*

THE METOPROLOL ATHEROSCLEROSIS PREVENTION IN HYPERTENSIVES (MAPHY) TRIAL

In 1987 the results of the Heart Attack Primary Prevention in Hypertension (HAPPHY) trial was published [487]. 6569 hypertensive men received either a diuretic (3272) or a beta-blocker (3297) for an average of 45.1 months. The result of the trial was that mortality was similar in the diuretic and beta-blocker groups. The interpretation of this trial was clear and it has not been seriously criticised. However the centres could opt to treat their diuretic treated patients with one of two diuretics and similarly one of two beta-blockers, atenolol or metoprolol. Unfortunately those centres who had chosen metoprolol continued to follow their patients for a further 14 months and reported their data as the results of the MAPHY trial [488].

The patients in the MAPHY trial had been part of the HAPPHY trial and contributed a large number of patient years. For the MAPHY trial, however, the end points were *"in part redefined, collected separately, and judged by separate, independent end point committees"* [489].

Thus the MAPHY trial was, in some respects, a subgroup analysis of the HAPPHY trial, in which 3234 men were followed for an average of 60 months. This analysis revealed an advantage for metoprolol over diuretic in terms of total, coronary heart disease and stroke mortality. The MRC trials, on the other hand, had tended to show an advantage for diuretics over beta-blockers [181,328].

The criticisms of this MAPHY 'subgroup analysis' were as follows:

i)   at the end of the trial mortality did not actually differ between the groups [490].
     Answer: a test should be applied that analyses the whole survival experience [491].

ii) *"it seems inescapable that in the atenolol group that was deleted from the MAPHY study the mortality rate must have been significantly higher.."* [492].
     Answer: *"the MAPHY and HAPPHY studies are two separate trials"* [491].

iii) *"it seems highly unlikely that somebody was not privy to the unblinded data of the HAPPHY trial when the MAPHY study was being conceived.*
     Is the MAPHY study the legitimate or illegitimate son of the HAPPHY trial?" [492].

In essence the MAPHY trial is a subgroup of the HAPPHY trial and the selection of a subpopulation raises questions over the results. Moreover when health economists have evaluated the cost-effectiveness of beta-blockers they have selected the MAPHY 'subgroup' with favourable results [493]. Such beneficial results would not be found by an analysis of the HAPPHY trial [422].

EXTRACRANIAL/INTRACRANIAL (EC/IC) BYPASS STUDY GROUP TRIAL

This trial randomised 1377 patients with symptomatic extracranial vascular disease to extracranial-intracranial bypass surgery or no surgery [494]. The trial showed no benefit from the surgical procedure over conventional management in those entered into the trial. However, Dudley was concerned that the result could not be extrapolated to the whole population with symptomatic extracranial vascular disease [495]. As patients could not be randomised if they were medically ineligible or if consent could not be obtained from the referring doctor or the patient, many patients could not enter the trial. This is, of course, the situation with most trials and that is why a log book should be kept to document all subjects considered for a trial (chapter 10). The trial result is only generalisable to those subjects similar to those who were randomised. With the EC/IC Bypass trial it appears that 1695 were operated on outside the trial at a time when 1077 were randomised [495,496].

In view of this finding Dudley wrote his editorial 'Extracranial-intracranial bypass, one; clinical trials, nil' [495] and stated:

"To exclude this proportion is serious cause for concern and leads to an initial reaction that the trial was biased. Trial researchers have two approaches to try to overcome this source of bias. Firstly, randomisation may be undertaken before consent is sought, and the subgroup that is excluded by failure to obtain consent can then be analysed not only for its standard characteristics but also for the proportions that would have entered each treatment arm. Secondly, even if this is not done, excluded patients must have their treatment recorded and be followed up with the same rigorous standards that pertain to those who have been included. The study of extracranial-intracranial bypass failed to take any of these precautions, and therefore selection bias by doctors, patients, or both cannot be excluded. Bias must therefore be assumed."

Fortunately bias does not have to be assumed. Only a small proportion of all patients are used in controlled trials [497] and we cannot evaluate the effect of treatment in those who are unfit for, or refuse, surgery. The results of the trial are expected to be generalisable to many patients considered for this operation.

REASONS FOR NONACCEPTANCE OF THE RESULTS OF RANDOMISED CONTROLLED TRIALS

The reasons for nonacceptance include: results at variance with preconceived ideas; errors in performance or design of the trial; errors in analysis; an atypical selection of patients; failure of randomisation to produce equivalent groups; failure to recruit enough patients;

faulty interpretation of results; inconsistency of the results within different groups in the same trial and between different trials; the adverse effects of treatment; and the fact that the trial originates from a group with a vested interest in a particular result.

Preconceived ideas not in agreement with the results

The Anturane Reinfarction Trial claimed that Anturane produced a large reduction in sudden deaths after myocardial infarction. This was not expected by most medical practitioners. A reduction was anticipated only for recurrent myocardial infarctions, and the unexpected result was one reason why the result of the Anturane Reinfarction Trial has not been accepted.

The European trial on streptokinase treatment showed a reduction in late mortality. This was unexpected and the trial criticised. Most importantly tolbutamide remains widely prescribed and the adverse effects in the UGDP trial have not been accepted.

Errors in the performance of the trial

Feinstein has criticised the University Group Diabetes Program trial for failing to define terms such as *congestive heart failure*; for having vague selection criteria; for failure to obtain important baseline data and information on the quality of life during the trial; for the quantity of missing data; for the difficulties in standardising the protocol (four clinics initially employed serum rather than whole blood determinations of glucose); and for discontinuing the use of tolbutamide before stopping other treatments [483].

Errors in the analysis of the trial

Errors in analysis were discussed in chapter 14. Three very important errors are emphasised here: failure to analyse on the intention-to-treat principle; the effects of repeated looks; and an inadequate classification of end points.

Failure to analyse on the intention-to-treat principle
The analysis of the Anturane Reinfarction Trial provides a classic example of the failure to analyse on the intention-to-treat principle. Sulphinpyrazone takes seven days to exert a full effect and the effect will be absent seven days after stopping the drug. The investigators therefore excluded patients in whom the drug could not have been active. They also excluded the corresponding placebo treated patients but discarded the safe intention-to-treat approach to analysis in favour of the per-protocol approach which may suffer from bias. Conventionally, for the main analysis, patients should be retained in their groups after randomisation, and this was the approach adopted by Elwood in his trials of aspirin following myocardial infarction [457,460].

Effect of repeated looks on the significance of the statistical tests reported
It appears that the problem of repeated looks was not considered initially in the Anturane Reinfarction Trial.

Classification of end points not well defined
The definition of sudden death in the Anturane Reinfarction Trial was inadequate and a reclassification of these deaths may have altered the conclusions to some extent.

Restricted selection of patients at entry

Patients entering a trial do not usually represent patients in general. This fact has led to criticism of many trials such as the EC/IC Bypass Study Group trial and the Anturane Reinfarction Trial. Many trials will fail to achieve general validity (chapter 13), but in the Anturane Reinfarction Trial, patients originally diagnosed as having a myocardial infarction were removed from the trial after randomisation, apparently distorting the results. However, the exclusion of patients before randomisation only reduces the general applicability of the results and dose not produce bias.

The European trial of streptokinase treatment was criticised as only 13.4 percent of available patients were entered into the trial. Most of the patients were excluded due to inability to give the treatment within 12 hours of the onset of chest pain. It was reasonable to exclude these patients from the trial as the treatment was only thought to exert an effect in the first 12 hours. It would be more sensible to recalculate the inclusion rate with the numbers presenting within 12 hours as the denominator.

Treatment groups not identical at entry

Randomisation worked effectively in the Anturane Reinfarction Trial to give similar groups, but this does not always occur. In the European trial of streptokinase and in the University Group Diabetes Program trial the groups were not identical in some important respects.

Too few patients in the trial

The small numbers of patients in many of the trials of beta-adrenoceptor blocking drugs has led to several being reported as negative but with very low power. Baber and Lewis [126] have graphically illustrated the low power in these trials by providing the confidence limits for the differences between treatments.

A trial may give a negative result not only if too few patients are entered but also if the true effect of treatment is very small. If a treatment confers only a small benefit, the trials have to be larger to prove this with any certainty. For example, if the true reduction in mortality with active treatment is greater than 50 percent, fewer patients will be required to prove this effect than to demonstrate a 20 percent reduction. In the secondary prevention of myocardial infarction, the effects of certain treatments may not be dramatic; for example, reductions in total mortality of up to 15 percent for aspirin and 20 percent for anticoagulants.

Faulty interpretation of the trial results

An example of faulty interpretation is provided by trials of antihypertensive agents where baseline blood pressure is determined before the start of the trial. As the patients become accustomed to the procedures adopted during the trial, their average blood pressure tends to fall. This order effect may thus be superimposed on any treatment effect, inflating the apparent effect of treatment. Douglas-Jones and Cruickshank [498] examined the use of atenolol as an antihypertensive agent and compared three different doses in a cross-over, random order, double-blind fashion. Unfortunately, baseline blood pressure was always determined prior to the commencement of the trial. The authors were able to conclude that there was no difference between the blood pressure on the three doses, but they should have been more cautious in concluding, *"Atenolol effectively decreased lying and standing blood pressures"* as the fall in pressure from the start of the trial may have been enhanced by giving the active doses last. The baseline assessment should have consisted of a double-blind period of placebo treatment given in random order during the body of the trial. However, the authors appear to have determined a correct baseline pressure, neither inflated by observer bias nor order effect, as the result of this trial agrees quantitatively with other trials where baseline blood pressure was determined correctly [499].

Trial results not consistent in different subgroups

It is desirable to examine the results of the trial in subgroups that are obviously important such as the two sexes, different races, and various age groupings. It is undesirable to invent subgroups after examining the data. For example, the best results may be observed in unmarried Chinese women over the age of 70 but the trial may include few such persons and a report on a small selected subgroup will be misleading.

After analysing the results in different groups, the effect of treatment may be shown to be inconsistent; this may raise doubts as to the generality of any conclusions. In an important trial of specialised care for hypertensive patients, the Hypertension Detection and Follow-up Program Trial, small improvements in blood-pressure control were associated with an overall decrease in mortality. The result was not found in white women when analysed separately [138,282]. This subgroup was not small and raises the question of the generality of the results. Similarly, it appears that men, but not women, may benefit from aspirin in the prevention of venous thromboembolism [500]. In the trials of beta-adrenoceptor blocking drugs, different effects have been reported in the elderly [477] that were not confirmed in other trials [218,479]. In one trial of these drugs, there was a better response with an anterior myocardial infarction [414] whereas in a second trial a greater treatment effect was observed with a posterior infarct [218]. Great caution has to be employed in subgroup analysis (chapter 14).

Different trials give different results

In the sections on trials in the secondary prevention of myocardial infarction (the use of aspirin; anticoagulants; beta-adrenoceptor blocking drugs) different results were apparent, some trials showing benefit and others none. It is a small wonder that the positive results were not widely accepted. However, combining the results can provide a more clear picture of any overall pattern [76].

The treatment is too difficult or has too many adverse effects

If a treatment is uniformly effective and capable of being provided, it will presumably be offered irrespective of difficulties of administrative and laboratory control. However, treatments are not uniformly effective; only a proportion of patients will benefit and the costs and difficulties have to be taken into account. The complexity of treatment was given as a reason for not implementing the results of the European Trial of Streptokinase in acute myocardial infarction.

The trials of anticoagulant treatment following myocardial infarction indicated a reduction in male mortality of 20 percent, but the complexity of treatment was such that its use has declined dramatically in most countries.

Vested interest of originating group

The Anturane Reinfarction Trial was criticised on account of being funded and analysed by the pharmaceutical company making Anturane, but this trial employed independent policy and audit committees and this criticism should not be taken seriously. However, pharmaceutical companies are responsible for a number of promotional trials and some of these require close attention. These trials are usually concerned with the acceptability of their products and a comparison of these with those of their competitors. The motivation for these trials arises from competitive marketing.

CONCLUSIONS

This chapter considered 11 reasons why the result of a particular trial may be rejected and gave several examples to support these assertions. As trial design, execution, and analysis improve it is hoped that the proportion of results that are rejected will be reduced. Results will still be falsely positive on occasions, and little can be done to reduce the strength of a reader's preconceived ideas, to avoid inconsistency of the treatment effect in different subgroups, or prevent different results emerging from separate trials. However, errors of performance, analysis, and interpretation can be minimised, sufficient representative patients can be recruited to ensure generalisability and limit the number of falsely negative trials, and trials sponsored by the pharmaceutical companies can be monitored and analysed by independent bodies. Even if a result is accepted it may not lead to any change in clinical practice. A treatment will not be employed if it is considered too difficult, has too many adverse effects, or is too expensive.

As the standard of randomised controlled trials improves, so will the quality of critical appraisal. We must admit that in the future the results of trials may be as frequently rejected as they are today.

20. THE ADVANTAGES AND DISADVANTAGES OF RANDOMISED CONTROLLED TRIALS

This chapter reviews the advantages and disadvantages of performing randomised controlled trials (RCTs). Observational studies may fail to include controls or may use an inappropriate comparison group whereas the RCT is, by definition, controlled with the best possible comparison group. This provides the main advantage of the randomised controlled trial — the greater confidence with which we can accept the results. Randomised controlled trials have prevented the introduction and continued use of useless and dangerous treatments. Employing current terminology *'RCTs provide the essential background to practising Evidence-Based medicine'*.

On the debit side, although we may have confidence in a result, the result may nevertheless be incorrect. Also, it is possible that the performance of a randomised controlled trial may delay the introduction of a useful treatment. Further, when a patient is randomised to the treatment that proves least effective, that individual may suffer as a consequence. Although the latter is true, a randomised controlled trial should only be used when there is genuine doubt about the efficacy of the treatments and the investigator cannot know if patients will suffer. A placebo may not prove to be the less desirable treatment as in some trials this has proven to be the most beneficial therapy [63,74]. Lastly, it is expensive to perform a randomised controlled trial and when the effect of treatment is very marked it may be unnecessary.

ADVANTAGES OF RANDOMISED CONTROLLED TRIALS

Randomised controlled trials provide the best estimate of the beneficial effects of treatment

The benefit to be derived from treatment must be established by comparing the results with either no treatment or some alternative therapy. Such control groups may be non-randomly allocated groups studied in the past (historical controls) or, less commonly, concurrently; or subjects included in a randomised controlled trial where randomisation provides a comparable control group that is treated simultaneously. Attempts to establish that historical controls are adequate [501] have not been supported by the literature [22,156,157]. Retrospective data are not valid for comparative purposes if the pattern of disease changes with time or the type of patient differs between two periods of observation. For example, an investigator may study a series of patients and then announce that he intends to evaluate a new treatment in the future. Patients that are referred to him for the new treatment may be more or less severely affected than the control group. However, a leading article concluded, *"If the change is rapid, ..., then a randomised trial may be invalidated just as much as any other"* [502]. This is difficult to accept as the control group will be treated simultaneously with the treatment under investigation and the result of a randomised controlled trial will be much less influenced by any change than a comparison with historical controls.

Randomised controlled trials may provide proof that a supposedly beneficial treatment is dangerous

One of the largest contributions to scientific knowledge resulting from the use of randomised controlled trials has been the discovery, not only that a treatment may be useless, but that it may be dangerous. This was discussed in chapter 16, and it cannot be stressed too strongly that even if a treatment appears to have a beneficial action in the short-term, the long-term effects must still be assessed in a randomised controlled trial. For example, clofibrate, a drug that lowers serum cholesterol, was compared with placebo in the primary prevention of ischaemic heart disease. Although serum cholesterol was lowered by treatment and some heart attacks were prevented, overall mortality was increased by this drug treatment – an unexpected finding that could not have been detected without a large randomised controlled trial.

DISADVANTAGES OF RANDOMISED CONTROLLED TRIAL

Falsely negative results

The most common example of a misleading answer is the false negative result. The term *misleading* is used rather than incorrect as a trial provides an estimate of a treatment effect and the authors and readers should consider the confidence limits for that result. A treatment effect may not be statistically significant but if the 95 percent confidence limits lie between a deterioration of 30 percent and an improvement of 50 percent, the conclusion of no proven effect may be misleading but not erroneous. Put another way, if a trial is too small to detect a true benefit, this is not a criticism of randomised controlled trials in general but only of a particular trial. Several small trials may sometimes be combined to give an overall estimate of the effect of a treatment (chapter 14). The methods for combining the results from many trials have been published and such analyses may be able to include the results from small trials.

Unfortunately, there may be a tendency not to publish the results from trials that prove negative. Both authors and editors will be inhibited from publication and Peto [161] and others [503] have observed that smaller trials tend to show greater positive effects than larger trials. Presumably a large trial, requiring more effort and providing a more dependable result, tends to be published whether negative or otherwise, whereas small trials may be published more frequently when a positive result is observed. Therefore Maxwell has suggested that journal editors *"assist the identification of negative results by publishing such work by title only - rather than not at all"* [504]. Such papers would have to be reviewed and registered with the editor even if they were not printed, so as to be available on request.

Falsely positive results

It appears rare for a trial to give a false positive result. At a symposium on clinical trials Sir Richard Doll identified one example [505] and Maxwell has quoted another [504]. Maxwell considered that with results achieving the 5 percent level of significance, one in 20 would <u>not</u> be expected to be in error as, *"Ethical considerations demand a sound scientific reason for believing that the null hypothesis [no difference between treatments] is certainly not likely to be true. Thus, in clinical research this error is much rarer than 5 percent and its detection even rarer still - also for ethical reasons"*. We can agree that a trial with a positive outcome in favour of a treatment may not be repeated for ethical reasons and therefore a false positive result may not be detected. Also promising treatments tend to be investigated, compared with no treatment and when the treatment is effective then, by definition, a false positive result will not be observed. Nevertheless many trials compare different active treatments but a high probability that one or more of 20 of trials will be falsely positive is not seen. Rarely is the exact 5% level of significance achieved and often considerably lower P values are reported. The probability is 36% of a false positive result if we found 20 trials, all with P=0.05 (chapter 6). A randomised controlled trial is no worse than any other method of investigation in rejecting the null hypothesis when it is correct and the absence of bias in well-conducted randomised controlled trials will greatly reduce the number of false positive results.

Delay of a valuable new treatments

An editorial in the *British Medical Journal* discussed early randomised controlled trials with a new treatment and stated [502], *"Our current insistence on randomised controlled trials has undoubtedly had a salutary effect on loose thinking but more than once this has been at the expense of progress."* Unfortunately, an example was not provided to substantiate the latter claim but it is theoretically possible that the performance of randomised controlled trials will delay the introduction of a new and effective treatment. Fortunately, an insistence on randomised controlled trials may also prevent progress with a useless or dangerous treatment.

Best treatment denied to control group

Fincke discussed a hypothetical trial of a new anticancer drug and concluded that if more patients die in the control group, then the doctors are guilty of manslaughter [507]. However, it can be argued that if a randomised controlled trial is not performed and a new treatment is not discovered to be associated with more deaths than the control treatment, and if the treatment is widely introduced, many doctors will be guilty of manslaughter and many more patients will be victims.

Ethical problems are avoided if there is genuine doubt about the efficacy of a new treatment. Also, when informed consent is sought the patient knows he may be randomised to a control group and gives his consent to take part in the trial. A leading article in the *Lancet* discussed the problems of not obtaining informed consent and quoted a double-blind placebo-controlled trial of the side effects of oral contraceptives. Six pregnancies occurred on placebo when the subjects were not aware that a placebo was being employed, although they were advised to use a spermicidal cream [508,509]. To avoid ethical problems, the subjects must be fully informed of the nature of the trial. This poses difficulties in paediatrics, psychiatry, and sometimes in the treatment of patients with cancer [510].

When a disease is nearly always fatal and no effective treatment is available, then a randomised controlled trial may be unnecessary since any improvement is obvious [508]. When the outcome is not always predictable a sequential trial or a trial with variable allocation of subjects to treatment may avoid some of the ethical pitfalls of withholding effective treatment from large numbers of patients (chapter 5).

Expense of randomised controlled trials

The performance of a randomised controlled trial is more expensive than uncontrolled observational studies. However, when a comparison group is required to demonstrate a treatment effect, historical controls are not adequate and a randomised controlled trial should be performed. The additional expense will be amply repaid by the improved quality of the data.

Undetectable small treatment effect

Randomised controlled trials do have their limitations and unless very large they will not detect small effects of treatment. A trial to detect a 5-10 percent reduction in a common event such as myocardial infarction will require a large number of subjects studied for a long period of time. A large trial may detect a 25 percent reduction in a common event but not an equivalent reduction in a rare event.

A rare adverse effect may not be detected even by a large randomised trial. Clinical impression, monitoring systems, and national vital statistics provide the only hope that a very rare adverse reaction will be discovered (chapter 16).

Effect obscured by several confounding factors

An outcome may be affected by many factors other than treatment. Black has argued that if attempts are made to restrict randomisation according to many factors, effective randomisation becomes impossible [511]. The answer is not to restrict randomisation for many factors but to allow for any difference retrospectively.

CONCLUSIONS

This chapter in itself is not intended to persuade the reader that the advantages of randomised controlled trials outweigh the disadvantages. Rather, the entire book is directed to this end. It may be fitting to conclude with Cochrane's early finding that, in the Northern Hemisphere, randomised controlled trials had failed to spread to the Catholic South and Communist East. Cochrane considered that the main explanation was the lower extent to which medical students were scientifically educated in these regions. However, he pointed out that Sweden, although situated in the North and West, was an exception. There were few randomised controlled trials carried out in Sweden compared to the number of meticulous observational studies performed. Cochrane admits that other factors may have influenced the performance of randomised controlled trials: the memory of war crimes in Germany, the extent of private practice and, the authoritarian structure of Soviet medicine [512]. Today Sweden, Italy and other countries that were slow to embrace clinical trials now have an excellent record of performing randomised trials.

The main advantage of randomised controlled trials lie in the confidence with which we can view the results; the disadvantages are trivial in comparison. I therefore hope that more randomised controlled trials will be performed in the future and their use extended much further in fields such as surgery, obstetrics, orthopaedics, psychiatry, physiotherapy, and sociology. This book is dedicated to any person who embarks on a randomised controlled trial as a result of reading it.

REFERENCES

1. Bradford Hill A. The clinical trial. Br Med Bull 1951;7:278-282.
2. Cochrane AL. Effectiveness and efficiency: Random reflections on health services. London. The Nuffield Provincial Hospital Trust, 1972.
3. Armitage P. Bradford Hill and the Randomized Controlled Trial. Pharmaceutical Medicine 1992;6:23-37.
4. Doll R, Peto R. Randomised controlled trials and retrospective controls. BMJ 1980;44:280.
5. Medical Research Council Investigation. Streptomycin treatment of pulmonary tuberculosis. BMJ 1948;2:769-782. Material reproduced with permission.
6. Medical Research Council. The prevention of whooping-cough by vaccination: a report of the Whooping-Cough Immunization Committee. BMJ 1951:1:1463-1471.
7. L'Etang JC. "Historical aspects of drug evaluation". In: *The principles and practice of clinical trials*. Harris EL and Fitzgerald JD (eds). E and S Livingstone, Edinburgh and London, 1970. Pp 3-6.
8. Bull JP. The historical development of clinical therapeutic trials. J Chronic Dis. 1959;10:218-248.
9. Bradford Hill A. Statistical methods in clinical and preventive medicine. Edinburgh and London, E and S Livingstone, 1962.
10. Maitland C. Account of inoculating the smallpox. London, J Robertson, 1972.
11. Williams W. Masters of medicine. London, Pan Books, 1954. P 23.
12. Jenner E. An inquiry into the cause and effects of the variolae vaccinae. London, S Low, 1798.
13. Pearson G. An inquiry concerning the history of cowpox. London, J Johnson, 1798.
14. Waterhouse B. A prospect of exterminating the smallpox. Boston, Cambridge, 1800.
15. Haygarth J. Of the imagination as a cause

388

and cure of disorders of the body. new ed. Bath, R Crutwell, 1801. P 3.

16. Sutton HG. Cases of rheumatic fever treated for the most part by mint water. Guy's Hosp Rep 1865;2:392-428.
17. Laplace PS. Théorie analytique des probabilités. Paris, 1812.
18. Louis PCA. Essay on clinical instruction. Translated by Martin P. London, S Highley, 1834.
19. Louis PCA. Recherches sur les effects de la saignée. Paris, De Mignaret, 1835.
20. Bartlett E. An essay on the philosophy of medical science. Philadelphia, Lea and Blanchard, 1844.
21. Lister J. On the effects of the antiseptic system upon the salubrity of a surgical hospital. Lancet 1870;1:4 and 40.
22. Pocock SJ. Randomised controlled trials [letter]. BMJ 1977;1:1661.
23. Vallery-Radot P. Pasteur 1822-1895. Paris, 1922.
24. Fibiger J. Om serumbehandling af Difteri. Hosp Tid, Kjøbenh 1898;4:6,309,337.
25. Pearson K. Report on certain enteric fever inoculation statistics. BMJ 1904;2:1243-6.
26. Medical Research Council. The serum treatment of lobar pneumonia: a report of the Therapeutic Trials Committee. BMJ 1934;1:241-5.
27. Porrit AE, Mitchell GAG. "An investigation into the prophylaxis and treatment of wound infection". In: Penicillin therapy and control in 21 army group. London, 21 Army Group, 1945. P7.
28. Peirce CS, Jastrow J. On small differences of sensation. National Academy of Sciences Memoirs 1884;3:75-83.
29. Fisher RA. The Design of Experiments. Edinburgh: Oliver & Boyd, 1935.
30. Amberson JB, McMahon BT, Pinner M. Clinical trials of sanocrysin in pulmonary tuberculosis. Am Rev Tuberc 1931;24:401-435.
31. Collins R, Julian D. British Heart Foundation surveys (1987 and 1989) of United Kingdom treatment policies for acute myocardial infarction. Br Heart J

1991;66:250-5.

32. Glaser EM. "Ethical aspects of clinical trials". In: *The principles and practice of clinical trials*. Harris EL and Fitzgerald JD (eds). Edinburgh and London, E and S Livingstone, 1970.

33. Wade OL. "Human experiment 2, clinical aspects". In: *Dictionary of medical ethics*. Duncan AS, Dunstan ER and Welbourn RB (eds). London, Darton Longman and Todd, 1977.

34. Mitscherlich A, Mielke F. Doctors of infamy: The story of the nazi medical crimes. Shuman, New York, 1949. Pp xxiii-xxv.

35. World Medical Association. Human experimentation, code of ethics of the World Medical Association. Declaration of Helsinki. BMJ 1964;2:177.

36. Bradford Hill A. Medical ethics and controlled trials. BMJ 1963;1:1043-1049. Material reproduced with permission of author and editors of BMJ.

37. Glover J. The MRC and informed consent. BMJ 1986;293:157-8.

38. King J. Informed consent [letter]. BMJ 1986;293:562.

39. Simes RJ, Tattersall MHN, Coates AS, Raghavan D, Solomon HJ, Smartt H. Randomised comparison of procedures for obtaining informed consent in clinical trials of treatment for cancer. BMJ 1986;293:1065-68.

40. Consent: How informed? Lancet 1984;1445-47.

41. Medical Research Council. Responsibility in investigations on human subjects. BMJ 1964;2:178-180.

42. Nicholson R. Informed consent [letter]. BMJ 1986;293:1099.

43. Committee Appointed by the Royal College of Physicians of London: Supervision of the Ethics of Clinical Investigations in Institutes. BMJ 1967;2:429-430. Later report published by the college: Report of committee on the supervision of the ethics of clinical investigation in institutions. 1973.

44. Wald N. Ethical issues in randomised prevention trials. BMJ 1993;306:563-5.

45. Hamblin TJ. A shocking American report with

lessons for all. BMJ 1987;295:73.

46. Report of a Committee Appointed by Governor Dwight H Green of Illinois. Ethics governing the service of prisoners as subjects in medical experiments. JAMA 1948;136:457-458.

47. Zelen M. A new design for randomized clinical trials. NEJM 1979;300:1242-1245.

48. Taylor KM, Margolese RG, Soskolne CL. Physicians' reasons for not entering eligible patients in a randomized clinical trial of surgery for breast cancer. NEJM 1984;310:1363-66.

49. Cancer Research Campaign Working Party in Breast Conservation. Informed consent: ethical, legal, and medical implications for doctors and patients who participate in randomised clinical trials. BMJ 1983;286:1117-1121.

50. Beecher HK. Ethics and clinical research. NEJM 1966;274:1354-1360.

51. Beecher HK. Pain, placebos and physicians. Practitioner 1962;189:141-155.

52. Veterans Administration Co-operative Study Group on Anti-Hypertensive Agents. Effects of treatment on morbidity in hypertension. II Results in patients with diastolic blood pressure averaging 90 through 114 mmHg. JAMA 1970;213:1143-1152.

53. Curson DA, Hirsch SR, Platt SD, Bamber RW, Barnes TRE. Does short term placebo treatment of chronic schizophrenia produce long term harm? BMJ 1986;293:726-728.

54. Hellman S, Hellman DS. Sounding Board of mice but not men: Problems of the randomized clinical trial. NEJM 1991;324:1585-89.

55. Beecher HK. Surgery as placebo: A quantitative study of bias. JAMA 1961;176:1102-1107.

56. Cobb LA et al. An evaluation of internal-mammary-artery ligation by a double-blind technique. NEJM 1959;260:1115-1118.

57. Dimond EG, Kittle CF, Crockett JE. Evaluation of internal-mammary artery ligation and sham procedure in angina pectoris. Circulation 1958;18:712-713.

58. Adams R. Internal-mammary-ligation for coronary insufficiency: An evaluation. NEJM

1958;258:113-115.

59. Fish RC, Grymes TP, Lovell MG. Internal-mammary-artery ligation for angina pectoris: Its failure to produce relief. NEJM 1958;259:418-420.

60. Amery A, Birkenhager W, Brixko P, Bulpitt CJ, Clement D, Deruyttere M, de Schaepdrijver A, Dollery CT, Fagard R, Forette F, Forte K, Hamdy R, Henry JF, Joossens JV, Leonetti G, Lund-Johansen P, O'Malley K, Petrie J, Strasser T, Tuomilehto J, Williams B. Mortality and morbidity results from the European Working Party on High Blood Pressure in the Elderly. Lancet, 1985;i:1349-1354.

61. McPherson K. Statistics: The problem of examining accumulating data more than once. NEJM 1974;290:501-502.

62. Coronary Drug Project Research Group: The Coronary Drug Project. JAMA 1970;214:1303-1313.

63. University Group Diabetes Program: A study of the effects of hypoglycaemic agents on vascular complications in patients with adult-onset diabetes. Diabetes 1970; 18(suppl2):747-830.

64. Veterans Administration Co-operative Study Group on Anti-hypertensive Agents: Effects of treatment on morbidity in hypertension. Results in patients with diastolic blood pressure averaging 115 through 129 mmHg. JAMA 1967;202:1028-1034.

65. The Anturane Reinfarction Trial Research Group: Sulfinpyrazone in the prevention of cardiac death after myocardial infarction. NEJM 1978;298:289-295.

66. Merigan TC. You can teach an old dog new tricks - how AIDS trials are pioneering new strategies. NEJM 1990;323:1341-3.

67. Waldron HA, Cookson RF. Avoiding the pitfalls of sponsored multicentre research in general practice. BMJ 1993;307:1331-4.

68. Royal College of Physicians. Guidelines on the practice of ethics committees in medical research involving human subjects. London: RCP, 1990.

69. Herxheimer A. Publishing the results of

392

sponsored clinical research: They are public, not private, property. BMJ 1993;307:1296-7.

70. Chalmers I. Underreporting research is scientific misconduct. JAMA 1990;263:1405-8.

71. Making clinical trialists register. Lancet 1991;338:244.

72. Freestone DS, Mitchell H. Inappropriate publication of trial results and potential for allegations of illegal share dealing. BMJ 1993;306:1112-4.

73. Bulpitt CJ, Semmence A, Whitehead T. Blood pressure and biochemical risk factors. Acta Cardiol 1978;33:109-110.

74. Committee of Principal Investigators: A Cooperative trial in the primary prevention of ischaemic heart disease using clofibrate. Br Heart J 1978;40:1069-1118.

75. Elwood PC, Cochrane AL, Burr ML, Sweetnam PM, Williams G, Welsby E, Hughes SJ, Renton R. A randomised controlled trial of acetyl salicylic acid in the secondary prevention of mortality from myocardial infarction. BMJ 1974;1:436-440.

76. An International Anticoagulant Review Group. Collaborative analysis of long-term anticoagulant administration after acute myocardial infarction. Lancet 1970;1:203-209.

77. Second International Study of Infarct Survival (ISIS-2) Collaborative Group. Randomised trial of intravenous streptokinase, oral aspirin, both, or neither among 17187 cases of suspected acute myocardial infarction: ISIS-2. Lancet 1988;ii:349-60.

78. GISSI. Effectiveness of intravenous thrombolytic treatment in acute myocardial infarction. Lancet 1986;i:398-402.

79. ISIS Steering Committee. Intravenous streptokinase given within 0-4 hours of onset of myocardial infarction reduced mortality in ISIS-2. Lancet 1987;i:502.

80. GISSI. Long-term effects of intravenous thrombolysis in acute myocardial infarction: final report of the GISSI study. Lancet 1987;ii:871-74.

81. AIMS Trial Study Group. Effect of thrombolysis in acute myocardial infarction: preliminary report of a placebo-controlled clinical trial. Lancet 1988;i:545-49.

82. Wilcox RG, vonder Lippe G, Olsson CG, Jensen G, Skene AM, Hampton JR. Trial of tissue plasminogen activator for mortality reduction in acute myocardial infarction: Anglo-Scandinavian Study of Early Thrombolysis (ASSET). Lancet 1988;ii:525-30.

83. Ketley D, Woods KL. Impact of clinical trials on clinical practice: example of thrombolysis for acute myocardial infarction. Lancet 1993;342:891-94.

84. A progress report of the European Working Party on High Blood Pressure in the Elderly (EWPHE). "Cardiac and Renal Function with increasing Age in Elderly Hypertensives". In: *Mild hypertension: Natural history and managements*. Gross F, Strasser T, eds. Pitman Medical, Tunbridge Wells. 1979. Pp181-197.

85. Amery A, Berthaux P, Bulpitt C, Deruyttere M, de Schaepdryver A, Dollery C, Fagard R, Forette F, Hellemans J, Lund-Johansen P, Mutsers A, Tuomilehto J. Glucose intolerance during diuretic therapy: Results of trial by the European Working Party on Hypertension in the Elderly. Lancet 1978;1:681-683.

86. Hills M, Armitage P. The two-period cross-over clinical trial. Br J Clin Pharmacol 1979;8:7-20.

87. Meier P, Free SM. Further consideration of methodology in studies of pain relief. Biometrics 1961;27:576-583.

88. Fisher RA. *The design of experiments*. Oliver and Boyd, Edinburgh, 1947.

89. Wilson C, Pollock MR, Harris AD. Diet in the treatment of infective hepatitis. Therapeutic trial of cysteine and variation of fat-content. Lancet 1946;1:881-883.

90. Aenishänslin W, Pestalozzi-Kerpel J, Dubach UC, Imhof PR, Turri M. Antihypertensive therapy with adrenergic beta-receptor blockers and vasodilators. Eur J Clin Pharmacol 1972;4:177-181.

91. Pearson RM, Bulpitt CJ, Havard CWH.

Biochemical and haematological changes induced by tienilic acid combined with propranolol in essential hypertension. Lancet 1979;1:697-699.

92. Pearson RM, Bending MR, Bulpitt CJ, George CF, Hole DR, Williams FM, Breckenridge AM. Trial of combination of guanethidine and oxprenolol in hypertension. BMJ 1976;1:933-936.

93. Chalmers J, Tillers D, Horvath J, Bune A. Effects of timolol and hydrochlorothiazide on blood-pressure and plasma renin activity. Double-blind factorial trial. Lancet 1976;2:328-331.

94. Lynch P, Dargie H, Krikler S, Krikler D. Objective assessment of anti-anginal treatment: a double-blind comparison of propranolol, nifedipine and their combination. BMJ 1980;1:184-187.

95. Williams EF. Experimental designs balanced for the estimation of residual effects of treatments. Aust J Sci Res Assoc 1949;2:149-168.

96. Cochran WG, Cox GM. In: *Experimental designs*. 2nd ed. Wiley & Sons, New York, 1957. P.133.

97. Armitage P. *Sequential modical trials*. Blackwell Scientific Publications, Oxford, 1960.

98. Wald A. Sequential analysis. Wiley, New York, 1947.

99. Robertson JD, Armitage P. Comparison of two hypotensive agents. Anaesthesia 1959;14:53-64.

100. Snell ES, Armitage P. Clinical comparison of diamorphine and pholcodine as cough suppressants, by a new method of sequential analysis. Lancet 1957;1:860-862.

101. Anscombe FJ. Fixed sample-size analysis of sequential observations. Biometrics 1954;10:98-100.

102. Cochran WG. "Newer statistical methods". In: *Quantitated methods in human pharmacology and therapeutics*. Lawrence DR (Ed.). Pergamon, London, 1959. Pp. 119-143.

103. Zelen M. Play the winner rule and the controlled trial. J Am Stat Assoc

1969;64:131-146.

104. Meier P. Terminating a trial - the ethical problem. Clin Pharmacol Ther 1979;25:633-640.

105. Chalmers TC. When and how to stop a clinical trial: Invited remarks. Clin Pharmacol Ther 1979;25:649-650.

106. Reiertsen O, Larsen S, Størkson R, et al. Safety of enoxaparin and dextran 70 in the prevention of venous thromboembolism in digestive surgery. A Play-the-Winner designed study. Scand J Gastroenterol 1993;28:1015-1020.

107. Larsen S, Reiertsen O, Mowinckel P, Lund H, Osnes M. Play the winner: a step-by-step process design to compare and control the efficacy and safety of treatments. Pharmaceutical Medicine 1994;8:11-23.

108. Byar DP, Simon RM, Friedewald WT, Schlesselman JJ, DeMets DL, Ellenberg JH, Gail MH, Ware JH. Randomised clinical trials. Perspectives on some recent ideas. NEJM 1976;295:74-80.

109. Hill C, Sancho-Garnier H. The two-armed bandit problem, a decision theory approach to clinical trials. Biomedicine 1978;28:42-43.

110. Guyatt GH, Keller JL, Jaeschke R, Rosenbloom D, Adachi JP, Newhouse MT. The n-of-1 randomised controlled trial: clinical usefulness, our three-year experience. Ann Intern Med 1990;112:293-9.

111. Guyatt G, Sackett D, Adachi J, Roberts R, Chong J, Rosenbloom D, Keller J. A clinician's guide for conducting randomised trials in individual patients. Can Med Assoc J 1988;139:497-503.

112. Jaeschke R, Adachi J, Guyatt G, Keller J, Wong B. Clinical usefulness of amitriptyline in fibromyalgia: the results of 23 n-of-1 randomised controlled trials. J Rheumatol 1991;18:447-51.

113. March L, Irwig L, Schwarz J, Simpson J, Chock C, Brooks P. n Of 1 trials comparing a non-steroidal anti-inflammatory drug with paracetamol in osteoarthritis. BMJ 1994;309:1041-6.

396

114. Campbell MJ. Commentary: n of 1 trials may be useful for informed decision making. BMJ 1994;309:1045-46,
115. Johannessen J. Controlled trials in single subjects. I. Value in clinical medicine. BMJ 1991;202:173-4.
116. Lewis JA. Controlled trials in single subjects. II. Limitation of use. BMJ 1991;303:175-6.
117. Senn SJ. Suspended judgement: n-of-1 trials. Control Clin Trials 1993;14:1-5.
118. Schwartz D, Flamant R, Lellouch J. Clinical Trials. Translated by MJR Healy. Academic Press, London, 1980.Pp 29-33.
119. Goodman SN, Berlin JA. The use of predicted confidence intervals when planning experiments and the misuse of power when interpreting results. Ann Intern Med 1994;121:200-206.
120. Hamilton M, Thompson EM, Wisniewski TK. The role of blood pressure control in preventing complications of hypertension. Lancet 1964;1:235-238.
121. Report of a co-operative randomised controlled trial. Control of moderately raised blood pressure. BMJ 1973;3:434-436. Material reproduced with permission.
122. Clark CJ, Downie CC. A method for the rapid determination of the number of patients to include in a controlled clinical trial. Lancet 1966;2:1357-1358.
123. National Diet-Heart Study Report. Appendix Aa-c. Sample size estimates for medical trials. Circulation 1968;suppl 37 and 38:1279-1308.
124. Braitman LE. Confidence intervals assess both clinical significance and statistical significance [Editorial]. Ann Intern Med 1991;114:515-7.
125. Rose GA. Beta-blockers in immediate treatment of myocardial infarction. BMJ 1980;208:1088.
126. Baber NS, Lewis JA. Beta-blockers in the treatment of myocardial infarction (letter). BMJ 1980;3:59.
127. Schwartz D, Lellouch J. Explanatory and pragmatic attitudes in therapeutical trials.

J Chron Dis. 1967;20:637-648.

128. Halperin M, et al. Sample sizes for medical trials with special reference to long-term therapy. J Chron Dis 1968;21:13-24.

129. George SL, Desu MM. Planning the size and duration of a clinical trial studying the time to some critical event. J Chron Dis 1974;27:15-24.

130. McPherson K. Statistics. The problem of examining accumulating data more than once. NEJM 1974;290:501-502.

131. Sondik EJ, Brown BW, Silvers A. High risk subjects and the cost of large field trials. J Chron Dis 1974;27:177-187.

132. Peto R, et al. Design and analysis of randomised clinical trial requiring prolonged observation of each patient. I. Introduction and design. Br J Cancer 1976;34:585-612.

133. Cochran WG. Sampling Techniques. Wiley, New York, 1963. Pp145.

134. Nam JM. Optimum sample sizes for the comparison of the control treatment. Biometrics 1973;29:101-108.

135. Gail M, et al. How many controls? J Chron Dis 1976;29:723-731.

136. Coronary Drug Project Research Group. The Coronary Drug Project: Design, methods and baseline results. Circulation 1973;suppl 47 and 48:12-137.

137. Altman DG. Size of clinical trials. BMJ 1983;286:1842-1843.

138. Hypertension Detection and Follow-up Program Cooperative Group. Five-year findings for the hypertension detection and follow-up program. I. Reduction in mortality of persons with high blood pressure including mild hypertension. JAMA 1979;242:2562-2571.

139. Roethlisberges FG, Dickson WJ. Management and the worker. An account of a research program conducted by the Western Electric Company, Hawthorne Works, Chicago. Harvard University Press, Cambridge, Mass., 1946.

140. Ederer F. Patient bias, investigator bias and the double-masked procedure in clinical trials. Am J Med 1975;58:295-299.

141. Beecher HK. The powerful placebo. JAMA

1995;159:1602-1606.

142. Shapiro AK. Factors contributing to the placebo effect. Their implications for psychotherapy. Am J Psychother 1964;18(suppl 1):73-88.

143. Keats AS, Beecher HK, Mosteller FC. Measurement of pathological pain in distinction to experimental pain. J Appl Physiol 1950;3:34-44.

144. Keats AS, D'Alessandro GL, Beecher HK. Report to the council on pharmacy and chemistry. JAMA 1951;147:1761-1776.

145. Beecher HK, et al. Field use of methodone and levo-iso-methadone in a combat zone. US Armed Forces Med J 1951;2:1269-1276.

146. Lasagne L, et al. Study of placebo response. Am J Med 1954;16:770-779.

147. Beecher HK, et al. Effectiveness of oral analgesics (morphine, codeine, acetylsalicylic acid) and problem of placebo 'reactors' and 'non reactors'. J Pharmacol Exp Ther 1953;109:393-400.

148. Travell J, et al. Comparison of effects of alpha-tocopherol and matching placebo on chest pain in patients with heart disease. Ann NY Acad Sci 1949;52:345-353.

149. Evans W, Hoyle C. Comparative value of drugs used in continuous treatment of angina pectoris. Q J Med 1933;2:311-338.

150. Greiner T, et al. Method for evaluation of effects of drugs on cardiac pain in patients with angina of effort; study of khellin (visamin). Am J Med 1950;9:143-155.

151. Bulpitt CJ. Heparin as an analgesic in myocardial infarction. A double-blind trial. BMJ 1967;3:279-281.

152. Jellinek EM. Clinical tests on comparative effectiveness of analgesic drugs. Biomet Bull 1946;2:87-91.

153. Gravenstein JS, Devloo RA, Beecher HK. Effect of antitussive agents on experimental and pathological cough in man. J Appl Physiol 1954;7:119-139.

154. Jillis BR. The assessment of cough-suppressing drugs. Lancet 1952;1:1230-1235.

155. Dollery CT. A bleak outlook for placebos (and for science). Eur J Clin Pharmacol

1979;15:219-221.

156. Rose G. Bias. Br J Clin Pharmacol 1982;13:157-162.

157. Christie D. Before-and-after comparisons: a cautionary role. BMJ 1979;2:1629-1630.

158. Zelen M. The randomisation and stratification of patients to clinical trials. J Chron Dis 1974;27:365-375.

159. Wright IS, Marple CD, Beck DF. Myocardial Infarction. Its clinical manifestation and treatment with anticoagulants. Grune and Stratton, New York, 1954. Pp 8-10.

160. Weinstein MC. Allocation of subjects in medical experiments. NEJM 1974;291:1278-1285. (Material reproduced with permission)

161. Peto R. Clinical trial methodology. Biomedicine Special 1978:24-36.

162. WHO (World Health Organisation) European Collaborative Group. European collaborative trial of multifactorial prevention of coronary heart disease; final report on the 6 year results. Lancet 1986;i:869-872.

163. Doll R, Peto R. Mortality in relation to smoking: 20 years' observation on male British doctors. BMJ 1976;2:1525-1536.

164. Rose G, Hamilton PJS. A randomised controlled trial of the effect on middle-aged men of advice to stop smoking. J Epidemiol Community Health 1978;32:275-281.

165. Ballintine EJ. Randomised controlled clinical trial. National Eye Institute workshop for ophthalmologists. Objective measurements and the double-masked procedure. Am J Ophthalmol 1975;79:763-767.

166. Pozdena RF. Versuche über Blondlots 'Emission Pesante'. Ann Physik 1905;17:104.

167. Seabrook W. Doctor Wood. Harcourt, Brace, New York 1941. Pp234.

168. Fletcher CM, "Criteria for diagnosis and assessment in clinical trials". In: Controlled clinical trials. AB Hill, ed. Springfield III. Charles C Thomas, 1960.

169. Kahn HA, et al. Serum cholesterol: its distribution and association with dietary and other variables in a survey of 10,000 men. Isr J Med Sci 1969;5:1117-1127.

170. Wilson EB. An introduction to scientific

research. McGraw-Hill, New York, 1952.

171. Bearman JE, Loewenson RB, Gullen WH. Muench's postulates, laws and corollaries or biometricians' views on clinical studies. Biometrics note 4, National Eye Institute, Bethesda, Md., 1974.

172. Foulds GA. Clinical research in psychiatry. J Ment Sci 1958;104:259.

173. Bulpitt CJ, Daymond M, Bulpitt PF, Ferrier G, Harrison R, Lewis PJ, Dollery CT. Is low salt dietary advice a useful therapy in hypertensive patients with poorly controlled blood pressure? Ann Clin Res 1985;16(suppl 43):143-9.

174. Knowelden J. In: *Prophylactic trials, medical surveys and clinical trials*. Witts LJ (Ed.). Oxford University Press, London. 1959.

175. National Diet-Heart Study Research Group. The National Diet-Heart Study. Circulation 1968;37(suppl 1):1253-59.

176. US Department of Health, Education and Welfare. National Institutes of Health. Cold study reveals some vitamin C influence; more research needed. Bethesda, Md. NIH Record 1973;25:4.

177. Heaton-Ward WA. Influence and suggestion in a clinical trial (Niamid in mongolism). J Ment Sci 1962;108:865-870.

178. Abraham HC et al. A controlled clinical trial of imipramine (Tofranil) with outpatients. Br J Psychiatry 1963;109:286-293.

179. Report to the Medical Research Council by its clinical psychiatry committee: Clinical trial of the treatment of depressive illness. BMJ 1965;1:881-886.

180. Report by the Management Committee. Initial results of the Australian Therapeutic Trial in mild hypertension. Clin Sci 1979;57:449s-452s.

181. Medical Research Council Working Party. MRC trial of treatment of mild hypertension: principal results. BMJ 1985;291:97-104.

182. SHEP Cooperative Research Group. Prevention of stroke by antihypertensive drug treatment in older persons with isolated systolic

hypertension: final results of the Systolic Hypertension in the Elderly Program (SHEP). JAMA 1991;265:3255-64.

183. Amery A, Birkenhager W, Bulpitt CJ, et al. Syst-Eur. A multicentre trial on the treatment of isolated systolic hypertension in the elderly: objectives, protocol, and organisation. Aging Clin Exp Res 1991;3:287-302.

184. Bulpitt CJ. A risk-benefit analysis for the treatment of hypertension. Postgrad Med J 1993;69:764-774.

185. Davey-Smith G, Egger M. Who benefits from medical interventions. BMJ 1994;308:72-74.

186. Staessen J, Fagard R, Amery A. Isolated systolic hypertension in the elderly: implications of SHEP for clinical practice and for the ongoing trials. J Hum Hypertens 1991;5:469-74.

187. Tabar L, Fagerberg CJG, Gad A et al. Reduction in mortality from breast cancer after mass screening with mammography. Lancet 1985;i:829-32.

188. Roberts MM, Alexander FE, Anderson TJ, et al. Edinburgh trial of screening for breast cancer: mortality at seven years. Lancet 1990;335:241-46.

189. Fletcher A, Spiegelhalter D, Staessen J, Thijs L, Bulpitt C. Implications for trials in progress of publication of positive results. Lancet 1993;342:653-57.

190. Bearman JE. Randomised controlled clinical trial. National Eye Institute Workshop for Ophthalmologists. Writing the protocol for a clinical trial. Am J Ophthalmol 1975;79: 775-778.

191. McFate Smith W. "Problems in long-term trials". In: Mild hypertension: Natural history and management. Gross F, Strasser T, eds. Pitman Medical, Tunbridge Wells. England. 1979:Pp 244-255.

192. Bulpitt CJ, Fletcher AE, Amery A, Coope J, Evans JG, Lightowlers S, O'Malley K, Palmer A, Potter J, Sever P, Staessen J, Swift C. The Hypertension in the Very Elderly Trial (HYVET): Rationale, methodology and comparison with previous trials. Drugs

Aging 1994;5:171-183.

193. Spriet A, Simon P. Questions à se pour vérifier un protocole d'essai thérapeutique avant d'en entreprendre l'exécution. Thérapie 1977;32:633-642.

194. Clinical Trials Unit, Department of Pharmacology and Therapeutics, London Hospital Medical College. Aide-mémoire for preparing clinical trial protocols. BMJ 1977;1:1323-1324.

195. Meier P. Terminating a trial - The ethical problem. Clin Pharmacol Ther 1979;25:633-640.

196. Stamler J. When and how to stop a clinical trial: Invited remarks. Clin Pharmacol Ther 1979;25:651-654.

197. Benedict GW. LRC Coronary Prevention Trial: Baltimore. Clin Pharmacol Ther 1979;25:685-687.

198. Schoenberger JA. Recruitment to the Coronary Drug Project and the Aspirin Myocardial Infarction Study. Clin Pharmacol Ther 1979;25:681-684.

199. Croke G. Recruitment for the National Co-operative Gallstone study. Clin Pharmacol Ther 1979;25:691-694.

200. Prout TE. Patient recruitment: Other examples of recruitment and solutions. Clin Pharmacol Ther 1979;25:695-696.

201. The First International Study of Infarct Survival (ISIS-1) Collaborative Group. Randomised trial of intravenous atenolol among 16027 cases of suspected acute myocardial infarction: ISIS-1. Lancet 1986;ii:57-66.

202. The Third International Study of Infarct Survival (ISIS-3) Collaborative Group. ISIS-3: a randomised comparison of streptokinase vs plasminogen activator vs anistreplase; and of aspirin plus heparin vs aspirin alone among 41299 cases of suspected acute myocardial infarction. Lancet 1992;339:753-70.

203. The GISSI 1 International Study. Lancet 1986:i;397-401.

204. Hamilton M. Computer programmes for the medical man: A solution. BMJ 1965;2:1048-

1050.

205. Wright P, Haybittle J. Design of forms for clinical trials (1), (2) and (3). BMJ 1979;2:529-530, 590-592, 650-651.

206. Bulpitt CJ, Dollery CT, Carne S. A symptom questionnaire for hypertensive patients. J Chron Dis, 1974;27:309-323.

207. Nicholls DP et al. Comparison of labetalol and propranolol in hypertension. Br J Clin Pharmacol 1980;9:233-237.

208. Survey Control Unit, Central Statistical Office. Ask a silly question. Government Statistical Service HMSO, 1976.

209. Bennett AE, Ritchie K. Questionnaires in medicine: A guide to their design and use. Nuff. Prov. Hosp. Trust, Oxford, 1975.

210. Spilker B, Schoenfelder J. Data collection forms in clinical trials. Raven Press, New York, 1991.

211. Bulpitt CJ, Shaw KM, Hodes C, Bloom A. The symptom patterns of treated diabetic patients. J Chron Dis, 1976;29:571-583.

212. Tinker MA. In: *Bases for effective reading.* Minnesota Press, Minneapolis, 1965.

213. Clark HH. Psychol Rev 1969;76:387.

214. International nonproprietary names (INN) for pharmaceutical substances: WHO, 1988.

215. Bloch A. Murphy's law and other reasons why things go wrong. Magnum books, Methuen Paperbacks, London, 1979.

216. Amery A, Berthaux P, Birkenhager W, Boel A, Brixko P, Bulpitt CJ, Clement D, de Padua F, Deruyterre M, de Schaepdryver A, Dollery CT, Fagard F, Forette F, Forte J, Henry JF, Hellemans H, Koistinen A, Laaser U, Lund-Johansen P, MacFarlane J, Miguel P, Mutsers A, Nissinen A, Ohm OT, Pelemans W, Suchett-Kaye AI, Tuomilehto J, Willems J, Willemse P. Antihypertensive therapy in patients above age 60 years: fourth interim report of EWPHE. Clin Sci Mol Med, 1978;55:263-270.

217. Hampton JR. Presentation and analysis of the results of clinical trials in cardiovascular disease. BMJ 1981;282:1371-1373.

218. Norwegian Multicentre Study Group. Timolol-induced reduction in mortality and reinfarction in patients surviving acute

myocardial infarction. NEJM 1981;301:801-807.

219. Wilkinson GN. Estimation of missing values for the analysis of incomplete data. Biometrics 1958;14:257-286.

220. Gordis L. "Conceptual and methodologic problems in measuring patient compliance". In: *Compliance in health care*. Haynes RL, Taylor RW, Sackett DL, eds. John Hopkins University Press, Baltimore and London. 1979:Pp 23-65.

221. Sackett DL. "A compliance practical for the busy practitioner". In: *Compliance in health care*. Haynes RL, Taylor RW, Sackett DL, eds. John Hopkins University Press, Baltimore and London. 1979:Pp 286-294.

222. Feinstein AR et al. A controlled study of three methods of prophylaxis against streptococcal infection in a population of rheumatic children. II. Results of the first three years of the study, including methods for evaluating the maintenance of oral prophylaxis. NEJM 1959;260:697-702.

223. Park LC, Lipman RS. A comparison of patient dosage deviation reports with pill counts. Psychoparmacologie 1964;6:299-302.

224. Gordis L, Markowitz M, Lilienfeld AM. The inaccuracy in using interviews to estimate patient reliability in taking medications at home. Med Care 1969;7:49-54.

225. Bulpitt CJ, Clifton P, Hoffbrand BI. Factors influencing over and under-consumption of anti-hypertensive drugs. Acta Int Pharmcodyn Ther 1980;suppl:243-250.

226. Mushlin AI, Appel FA. Diagnosing potential noncompliance. Physicians' ability in a behavioural dimension of medical care. Arch Intern Med 1977;137:318-321.

227. Roth HP, Caron HS, Hsi BP. Measuring intake of a prescribed medication: A bottle count and a tracer technique compared. Clin Pharmacol Ther 1970;11:288-337.

228. Bergman AB, Werner RJ. Failure of children to receive penicillin by mouth. NEJM 1963;268:1334-1338.

229. Report of Medical Research Council Working Party on Mild to Moderate Hypertension.

Randomised controlled trial of treatment for mild hypertension: design and pilot trial. BMJ 1977;1:1437-1440.

230. Averbuch M, Weintraub M, Pollock DJ. Compliance monitoring in clinical trials: the MEMS device. Clin Pharmacol Ther 1988;43:185.

231. Urquhart J, Chevalley C. Impact of unrecognised dosing errors on the cost and effectiveness of pharmaceuticals. Drug Information J 1988;22:363-378.

232. Cramer JA, Mattson RH, Prevey ML, Scheyer RD, Ouellette VL. How often is medication taken as prescribed? A novel assessment technique. JAMA 1989;261:3273-3277.

233. Urquhart J. "Noncompliance: the ultimate absorption barrier". In: *Novel Drug Delivery and Its Therapeutic Application*. Prescott LF, Nimmo WS, eds. John Wiley & Sons, 1989:Pp127-137.

234. Joyce CRB. Patient co-operation and the sensitivity of clincal trials. J Chron Dis 1962;15:1025-1036.

235. Goldsmith CH. "The effect of compliance distributions on therapeutic trials". In: *Compliance in health care*. Haynes RL, Taylor RW, Sackett DL, eds. John Hopkins University Press, Baltimore and London, 1979;Pp297-308.

236. Feinstein AR. Biostatistical problems in 'compliance bias'. Clin Pharmacol Ther 1974;16:846-857.

237. General Medical Council: Disciplinary Committee. Erasures from Register. BMJ 1975;3:391-2.

238. Anonymous. GMC professional conduct committee. BMJ 1988;296:306.

239. Lock SP. Research fraud: discouraging the others. BMJ 1990;301:1348.

240. Wells F. Fraud and misconduct in clinical research: is it prejudicial to patient safety? Adverse Drug React Toxicol Rev 1992;11:241-255.

241. Good Clinical Practice in Europe. Allen MR, ed. Rostrum Publications, Essex, 1991.

242. Food and Drug Administration. Compliance Program Guidance Manual, 8348-811, Clinical Investigators, November 1988.

243. Good Clinical Practice for Trials on Medicinal Products in the European Community. III/3976/88-EN Final. Pharmacol Toxicol 1990;67:361-372.

244. Hutchinson DR. A practical guide to GCP for investigators. Brookwood Medical Publications Ltd, Surrey, 1993.

245. A practical guide to FDA GCP for investigators. ISBN 1-874409-70-6.

246. Palmer E, Hardy RJ, Davis BR, Stein JA, Mowery RL, Tung B, Phelps DL, Schaffer DB, Flynn JT, Phillips CL, on behalf of the CRYO-ROP Cooperative Group: Operational aspects of early termination in the Multicenter Trial of Cryotherapy for Retinopathy of Prematurity. Control Clin Trials 1991;12:277-292.

247. Davis BR, Wittes J, Pressel S, Berge KG, Morton Hawkins C, Lakatos E, Moyé LA, Probstfield JL. Statistical considerations in monitoring the Systolic Hypertension in the Elderly Program (SHEP). Control Clin Trials 1993;14:350-361.

248. Editorial. On stopping a trial before its time. Lancet 1993;342:1311-1312.

249. Volberding PA, Lagakos SA, Koch MA et al. Zidovudine in asymptomatic human immunodeficiency virus infection. NEJM 1990;322:941-49.

250. Pocock SJ. When to stop a clinical trial. BMJ 1992;305:235-40.

251. ISIS-3 Collaborative Group. ISIS-3 protocol; Third international study of infarct survival, Oxford: Clinical Trial Services Unit, 1989.

252. Errington RD, Ashby D, Gore SM, et al. high energy neutron treatment for pelvic cancers: study stopped because of increased mortality. BMJ 1991;302:1045-51.

253. Aboulker JP, Swart AM. Preliminary analysis of the Concorde trial. Lancet 1993;341:889-90.

254. Moertel CG, Fleming TR, Macdonald JS, et al. Levamisole and fluorouracil for adjuvant therapy of resected colon carcinoma. NEJM 1990;322:352-58.

255. Systolic Hypertension in the Elderly's

Collaborative Group Coordinating Centre. Systolic hypertension in the elderly: Chinese trial (Syst-China), interim report (in Chinese with English summary). Chin J Cardiol 1992;20:270-75.

256. Bulpitt CJ. Shall we treat isolated systolic hypertension in 1994? J Hum Hypertens 1994;8:785-788.

257. O'Brien PC, Fleming TR. A multiple testing procedure for clinical trials. Biometrics 1979;35:549-556.

258. The ß-Blocker Heart Attack Trial Research Group. A randomised trial of propranolol in patients with acute myocardial infarction. I. Mortality results. JAMA 1982;247:1707-14.

259. The ß-Blocker Heart Attack Trial Research Group. A randomised trial of propranolol in patients with acute myocardial infarction. II. Morbidity results. JAMA 1983;250:2814-9.

260. West of Scotland Coronary Prevention Study Group. A coronary primary prevention study of 6000 Scottish men aged 45-64 years: study design. J Clin Epidemiol 1992;45:849-860.

261. Dillman RO, Seagren SL, Propert K, Guerra J, Eaton WL, Perry MC et al. A randomised trial of induction chemotherapy plus high-dose radiation versus radiation alone in stage III non-small cell lung cancer. NEJM 1990;323:940-5.

262. Souhami RL, Spiro SG, Cullen M. Chemotherapy and radiation therapy as compared with radiation therapy in stage III non-small cell lung cancer. NEJM 1991;324:1136.

263. Cardiac Arrhythmia Suppression Trial Investigators. Preliminary report: effect of encainide and flecainide on mortality in a randomised trial of arrhythmia suppresion after myocardial infarction. NEJM 1989;321:406-12.

264. Cooper GR. "The World Health Organisation Centre for Disease Control Lipid Standardisation Program". In: *Quality Control in Chemistry*. Walter de Druter and Co, Berlin. 1976, pp 97-105.

265. Scott WA. Reliability of content analysis: The case of nominal scale coding. Public Opinion Quart 1955;19:321-325.

408

266. Cohen J. A coefficient of agreement for nominal scales. Educ Psychol Meas 1960;20:37-46.
267. Maxwell AE, Pilliner AEG. Deriving coefficients of reliability and agreement for ratings. Br J Math Stat Psychol 1968;21:105-116.
268. Fleiss JL. "The measurement of Interrater Agreement". In: *Statistical Methods for Rates and Proportions*. John Wiley and Sons, New York. 1981, chapter 13, pp 212-236.
269. Landis JR, Koch GG. The measurement of observer agreement for categorial data. Biometrics 1977;33:159-174.
270. Rogot E, Goldberg ID. A proposed index for measuring agreement in test-retest studies. J Chron Dis 1966;19:991-1006.
271. Bulpitt CJ, Dollery CT, Carne S. Change in symptoms of hypertensive patients after referral to hospital clinic. Br Heart J 1976;38:121-128.
272. Galton F. Regression towards mediocrity in hereditary stature. J Anthropological Inst 1886;15:246-263.
273. Ferris FL, Ederer F. External monitoring in multiclinic trials: Application from ophthalmologic studies. Clin Pharmacol Ther 1979;25:720-723.
274. Wright BM, Dore CF. A random-zero sphygmomanometer. Lancet 1970;1:337-338.
275. De Gaudemaris R, Folsom AR, Prineas RJ, Luepker RV. The random zero versus the standard mercury sphygmomanometer - a systematic blood pressure difference. Am J Epidemiol 1985;121:282-290.
276. Rose GA, Holland WW, Crowley FA. A sphygmomanometer for epidemiologists. Lancet 1964;1:296-300.
277. Fitzgerald D, O'Callaghan W, O'Malley K, O'Brien E, Inaccuracy of the London School of Hygiene Sphygmomanometer. BMJ 1982;284:18-19.
278. Kahn HA et al. Standardising diagnostic procedures. Am J Ophthalmol 1975;79:768-775.
279. Williams OD. A framework for the quality assurance of clinical data. Clin Pharmacol Ther 1979;25:700-702.

280. Fowler FG, Fowler HW. *The Pocket Oxford Dictionary of Current English*. SR ed. Oxford Clarendon Press. 1969.
281. Veterans Administration Co-operative Study Group on Antihypertensive Agents. Effects of treatment on morbidity in hypertension: II. Influence of age, diastolic pressure, and prior cardiovascular disease: Further analysis of side effects. Circulation 1972;45:991-1004.
282. Hypertension Detection and Follow-up Program Cooperative Group. Five-year findings of the hypertension detection and follow-up program: II. Mortality by race, sex and age. JAMA 1979;242:2572-77.
283. Moher D, Dulberg CS, Wells GA. Statistical power, sample size, and their reporting in randomised controlled trials. JAMA 1994;272:122-124.
284. Schor S, Karten I. Statistical evaluation of medical journal manuscipts. JAMA 1966;195:1123-1128.
285. Schoolman HM et al. Statistics in medical research: principles versus practices. J Lab Clin Med 1968;7:357-367.
286. Lionel NDW, Herxheimer A. Assessing reports of therapeutic trials. BMJ 1970;3:637-640.
287. Gore SM, Jones IG, Rytter EC. Misuse of statistical methods: critical assessment of articles in BMJ from January to March 1976. BMJ 1977;1:85-87.
288. Freiman JA, et al. The importance of beta, the type II error and sample size in the design and interpretation of the randomised control trial. Survey of 71 'negative' trials. NEJM 1978;299:690-694.
289. Glantz SA. Biostatistics: How to detect, correct and prevent errors in the medical literature. Circulation 1980;61:1-7.
290. Armitage P, Berry G. "*Statistical Methods in Medical Research*". Second edition. Blackwell Scientific Publications, London. 1987.
291. Everitt BS. *Statistical Methods for Medical Investigations*. Second Edition. Edward Arnold, London. 1994.
292. Bailar JC III, Mostellar F. *Medical Uses of Statistics*. Second edition. NEJM Books,

Boston, Massachusetts. 1992
293. Peto R, Pike MC, Armitage P, et al. Design and analysis of randomised clinical trials requiring prolonged observations of each patient. II: Analysis and examples. Br J Cancer 1977;35:1-39.
294. Hogben L, Sim M. The self-controlled and self-recorded clinical trial for low-grade morbidity. Br J Prev Soc Med 1953;7:163-179.
295. Bulpitt CJ. The design of clinical trials. Br J Hosp Med 1975;12:611-620.
296. Feinstein AR. Clinical biostatistics. A survey of the statistical procedures in general medical journals. Clin Pharmacol Ther 1974;
297. Zar JH. In: *Biostatistical analysis*. NJ Prentice-Hall, Englewood Cliffs. 1974, pp 130-131.
298. Godfrey K. "Comparing the means of several groups". In: *Medical Uses of Statistics*. Second edition. JC Bailar III, F Mostellar, eds. NEJM Books, Boston, Massachusetts. 1992, pp233-257.
299. Feller W. "An introduction to probability theory and its applications". Volume I (ed.3). Wiley, New York. 1968.
300. Altman DG. Comparability of randomised groups. Statistician 1985;34:125-136.
301. Chalmers TC, Celano P, Sacks HS, Smith H Jr. Bias in treatment assignment in controlled clinical trials. NEJM 1983;309:1358-1361.
302. Lavori PW, Louis TA, Bailar JC, Polansky M. Designs for experiments - parallel comparisons of treatment. NEJM 1983;309:1291-1298.
303. Nie NN, et al. *Statistical package for the social sciences*. McGraw-Hill, New York. 1975.
304. Healy MJR, Whitehead TP. Outlying values in the national quality control scheme. Ann Clin Biochem 1980;17:78-81.
305. Barnett V. The study of outliers: Purpose and model. Appl Stat 1978;27:242-250.
306. Sprackling ME, et al. Blood pressure reduction in the elderly: a randomised controlled trial of methyldopa. BMJ 1981;283:115101153.

307. Mantel N. Evaluation of survival data and two new rank order statistics arising in its consideration. Cancer Chemother Rep 1966;50:163-170.
308. Breslow NE. Analysis of survival data under the proportional hazards model. Int Stat Rev 1975;43:45-57.
309. Cox DR. Regression models and life-tables. J R Stat Soc B 1972;34:187-202.
310. Miller RG, Halpern J. Regression with censored data. Biometrika 1982;69:521-531.
311. Mantel N. Evaluation of survival data and two new rank order statistics arising in its consideration. Cancer Chemotherapy Reports 1966;50:163-170.
312. Bhattacharyya GK, Johnson RA. *Statistical concepts and methods*. Wiley, New York. 1977.
313. Bulpitt CJ. Confidence Intervals. Lancet, 1987;i:494-7.
314. Wulff HR. Confidence limits in evaluating controlled therapeutic trials (letter). Lancet 1973;2:969-970.
315. Leading article. Interpreting clinical trials. BMJ 1978;2:1318.
316. Gardner MJ, Altman DG. Confidence intervals rather than P values: estimation rather than hypothesis testing. BMJ 1986;292:746-50.
317. Baber NS, Lewis JA. Beta-blockers in treatment of myocardial infarction. BMJ 1980;2:59.
318. Katz D, Baptista J, Azer SP, Pike MC. Obtaining confidence intervals for the risk ratio in cohort studies. Biometrics 1978;34:469-71.
319. Diem K, Lentner C. *Documenta Geigy. Scientific tables*. 7th ed. Geigy, Basel. 1970.
320. Ederer F. A parametric estimate of the standard error of the survival rate. J Am Stat Assoc 1961;56:111-118.
321. Gore SM. Statistics in question: Assessing methods - confidence intervals. BMJ 1981;283:660-662.
322. Pocock SJ, Hughes MD. Estimation issues in clinical trials and overviews. Stat Med 1990;9:657-671.
323. McMichael J. Anticoagulants: Another view.

BMJ 1964;2:1007.
324. Bulpitt CJ. Subgroup Analysis. Lancet 1988:31-34.
325. Collins R, Gray R, Godwin J, Peto R. Avoidance of large biases and large random errors in the assessment of moderate treatment effects: the need for systematic overviews. Stats Med 1987;6:245-50.
326. Multicenter Diltiazem Postinfarction Trial Research Group. The effect of diltiazem on mortality and reinfarction after myocardial infarction. NEJM 1988;319:385-392.
327. Durrleman S, Simon R. Diltiazem and mortality and reinfarction after myocardial infarction (letter). NEJM 1989;320:123.
328. MRC Working Party. Medical Research Council trial of treatment of hypertension in older adults: principal results. BMJ 1992;304:405-16.
329. Coope J, Warrender TS. Randomised trial of treatment of hypertension in the elderly patients in primary care. BMJ 1986;293:1145-1148.
330. Dalhof B, Lindholm LH, Hansson L, Schersten B, Wester PO. Morbidity and mortality in the Swedish Trial in Old Patients with Hypertension (STOP-Hypertension). Lancet 1991;338:1281-5.
331. Insua JT, Sacks HS, Lau T-S, Lau J, Reitman D, Pagano S, Chalmers TC. Drug treatment of hypertension in the elderly: a meta-analysis. Ann Intern Med 1994;121:355-362.
332. Leonetti G, Cuspidi C, Fastidio M, Lonati L, Chianca R. Arterial hypertension as a risk factor in the elderly and its treatment. J Hypertens 1992;10:S3-7.
333. Celis H, Fagard R, Staessen J, Thijs L, Amery A. The older hypertensive. Assessment and treatment. Neth J Med 1993;43:S66-77.
334. Glass GV. Primary, secondary, and meta-analysis of research. Educ Res 1976;5:3-8.
335. Peto R. Why do we need systematic overviews of randomised trials? Stat Med 1987;6:233-40.
336. Furberg CT, Morgan TM. Lessons from overviews of cardiovascular trials. Stat Med 1987;6:295-303.

337. Sacks HS, Berrier J, Reitman D, Ancona-Berk VA, Chalmers TC. Meta-analyses of randomised controlled trials. NEJM 1987;316:150-55.

338. Editorial. Whither meta-analysis? Lancet 1987;i:897-98.

339. Bulpitt CJ, Meta-analysis. Lancet;93-94.

340. Bulpitt CJ. Does lowering blood pressure in hypertensive patients reduce the risk of death from coronary heart disease? Neth J Med 1987;30:228-34.

341. Wittes RE. Problems in the medical interpretation of overviews. Stat Med 1987;6:269-76.

342. Demets DL. Methods for combining randomised clinical trials: strengths and limitations. Stat Med 1987;6:341-48.

343. Lam W, Sacks HS, Sze PC, Chalmers TC. Meta-analysis of randomised controlled trials of nicotine chewing-gum. Lancet 1987;ii:27-29.

344. Mantel N, Haenszel W. Statistical aspects of the analysis of data from retrospective studies of disease. J Natl Cancer Inst 1959;22:719-48.

345. Magnesium, myocardial infarction, meta-analysis and megatrials. Drug and Therapeutic Bulletin 1995;33:25-27.

346. Teo KK, Yusuf S, Collins R, Held P, Peto R. Effects of intravenous magnesium in suspected acute myocardial infarction: overview of randomised controlled trials. BMJ 1991;303:1499-503.

347. Lau J, Antman EM, Jimenez-Silva J, Kupelnick B, Mosteller F, Chalmers TC. Cumulative meta-analysis of therapeutic trials for myocardial infarction. NEJM 1992;327:248-54.

348. Horner S. Efficacy of intravenous magnesium in acute myocardial infarction in reducing arrhythmias and mortality. Circulation 1992;86:774-9.

349. Woods KL, Fletcher S, Roffe C, Haider Y. Intravenous magnesium sulphate in suspected acute myocardial infarction: results of the second Leicester Intravenous Magnesium Intervention Trial (LIMIT-2). Lancet 1992;339:1553-8.

350. ISIS-4: a randomised factorial trial assessing early oral captopril, oral

414

mononitrate, and intravenous magnesium sulphate in 58,050 patients with suspected acute myocardial infarction. Lancet 1995;345:669-85.

351. Bulpitt CJ, Fletcher AE. Evaluation of side effects of antihypertensive treatment. High Blood Press 1993;2(suppl 1):70-75.

352. Amery A, et al. Hypotensive action and side effects of clonidine-chlorthalidone and methyldopa-chlorthalidone in treatment of hypertension. BMJ 1970;4:392-395.

353. McMahon FG. Efficacy of an antihypertensive agent. Comparison of methyldopa and hydrochlorothiazide in combination and singly. JAMA 1975;231:155-158.

354. Gibb WE, et al. Comparison of bethanidine, alpha-methyldopa and reserpine in essential hypertension. Lancet 1970;2:275-277.

355. Hefferman A, et al. A within-patient comparison of debrisoquine and methyldopa in hypertension. BMJ 1971;1:75-78.

356. Conolly ME, et al. A crossover comparison of clonidine and methyldopa in hypertension. Eur J Clin Pharmacol 1972;4:222-227.

357. Oates JA, et al. The relative efficacy of guanethidine, methyldopa and pargyline as antihypertensive agents. NEJM 1965;273:729-734.

358. Prichard BNC, et al. Bethanidine, guanethidine and methyldopa in treatment of hypertension: A within patient comparison. BMJ 1968;1:135-144.

359. Schooler KK. A study of errors and bias in coding responses to open end questions. Dissertation Abstr 1956;16:2542.

360. Young DW. Evaluation of a questionary. Methods Inf Med 1972;11:15-19.

361. Collen MF, et al. Reliability of a self-administered medical questionnaire. Arch Intern Med 1969;123:664-681.

362. Rose GA. The diagnosis of ischaemic heart pain and intermittent claudication in field surveys. Bull WHO 1962;27:645-658.

363. Maccoby EE, Maccoby N. "The interview: A tool of social science". In: *Handbook of social psychology*. Lindzey G ed. Addison-Wesley, Reading, Mass. 1954.

364. Mellner C. The self-administered medical history. Theoretical possibilities and practical limitations of the usefulness of standardised medical histories. Acta Chir Scand Suppl 1970;406:1.

365. Belson W, Duncan JA. A comparison of the check-list and the open response questioning systems. Appl Stat 1962;11:120-132.

366. Cannel CF, Kahn RL. "The collection of data by interviewing". In: *Research Methods in Behavioral Sciences*. Festinger L, Katz D, eds. Dryden, New York. 1953.

367. Cronbach LJ. Further evidence on response sets and test design. Educ Psych Meas 1950;10:3.

368. Anderson J, Day JL. New self-administered medical questionary. BMJ 1968;4:636-638.

369. Anastasi A. In: *Psychological testing*. Macmillan, New York. 1961.

370. Parten MB. In: *Surveys, polls and samples*. Harper, New York. 1950.

371. Shepherd M. "Implications of a multi-centred clinical trial of treatment of depressive illness". In: *Anti-depressant drugs*. Garattini S, Dukes MNG, eds. Exerpta Medica, Amsterdam. 1967. p332.

372. Hamilton M. "Evaluation of psychotropic drugs (3) sedatives." In: *Principles and practice of clinical trials*. Harris EL, Fitzgerald JD, eds. E & S Livingston, Edinburgh and London. 1970. pp 217-225.

373. Eysenck HJ. *Manual of the Maudsley personality inventory*. University of London Press, London. 1959

374. Taylor JA. A personality scale of manifest anxiety. J Abnorm Soc Psychol 1953;48:285.

375. Overall JE, Gorham GR. The brief psychiatric rating scale. Psychol Rep 1962;10:799.

376. Kellner R, Sheffield BF. A self-rating scale of distress. Psychol Med 1978;3:88-100.

377. Hamilton M. The assessment of anxiety by rating. Br J Med Psychol 1959;32:50.

378. Montgomery SA, Asberg M. A new depression scale designed to be sensitive to change. Br J Psychiat 1978;134:382-389.

379. Beck AT, Ward CH, Mendelson M, Mock J, Erbaugh J. An inventory for measuring

416

depression. Arch Gen Psychiat 1961;4:561-571.

380. Stevens B. Dependence of schizophrenic patients on elderly relations. Psychol Med 1972;2:17-32.

381. Shepherd M. Evaluation of psychotropic drugs." In: *Principles and practice of clinical trials*. Harris EL, Fitzgerald JD, eds. E & S Livingston, Edinburgh and London. 1970. pp 208-216.

382. Bullinger M. "Indices versus profiles – advantages and disadvantages". In: *Quality of Life Assessment*. Walker SR, Rosser RM, eds. Kluwer Academic Publishers, United Kingdom, 1993, pp. 209-220.

383. Bergner M, Bobbit RA, Carter WB, Gilson BS. The sickness impact profile. Development and final revision of a health status measure. Med Care 1981;19:787-805.

384. Hunt S. "The Nottingham Health Profile". In: *Assessment of Quality of Life in Clinical Trials of Cardiovascular Therapies*. Wenger NK, Mattson ME, Furberg CD, Ellinson J, eds. LeJacq Publishers, New York, 1984.

385. Kaplan RM, Bush JW, Berry CC. Health status: types of validity and the index of well-being. Health Serv Res 1976;11:478-507.

386. DyPuy HJ. "The Psychological Well-Being (PGWB) index". In: *Assessment of Quality of Life in Clinical Trials of Cardiovascular Therapies*. Wenger NK, Mattson ME, Furberg CD, Ellinson J, eds. LeJacq Publishers, New York, 1984, pp. 170-183.

387. McNair D, Lorr M, Droppelman LF. Manual for the Profile of Mood States. California, Educational and Industrial Testing Service, San Diego, 1971.

388. Zigmond A, Snaith P. The Hospital Anxiety and Depression Questionnaire. Acta Scand Psychiat 1983;67:361-368.

389. Bergner M. "Development, testing, and use of the Sickness Impact Profile". In: *Quality of Life Assessment*. Walker SR, Rosser RM, eds. Kluwer Academic Publishers, United Kingdom, 1993, pp. 95-110.

390. Rockey PH, Griep RJ. Behavioural dysfunction in hyperthyroidism. Improvement with

treatment. Arch Intern Med 1980;140:1194-1197.

391. Fletcher AE, McLoone P, Bulpitt CJ. Quality of life on angina therapy: A randomised controlled trial of transdermal glyceryl trinitrate against placebo. Lancet 1988;ii:4-8.

392. Parr G, Darekar B, Fletcher AE and Bulpitt CJ. Joint pain and quality of life; results of a randomised trial. Br J Clin Pharmac, 1988;27:235-242.

393. Fanshel S, Bush JW. A health status index and its application to health services outcomes. Operations Res 1970;18:1021-1065.

394. Bombardier C, Ware J, Russell IJ, et al. Auranofin therapy and quality of life for patients with rheumatoid arthritis: Results of a multicenter trial. Am J Med 1986;81:565-578.

395. Wu AW, Matthews WC, Brysk LT, Atkinson JH, Grant I, Abramson I, Kennedy CJ, McCutchan JA, Spector SA, Richman DD. Quality of life in a placebo-controlled trial of zidovudine in patients with AIDS and AIDS-related complex. J Acquired Immune Deficiency Syndromes 1990;3:683-690.

396. Visser MC, Fletcher AE, Parr G, Simpson A, Bulpitt CJ. A comparison of three quality of life instruments in subjects with angina pectoris: The Sickness Impact Profile, the Nottingham Health Profile, and the Quality of Well Being Scale. J Clin Epidemiol 1994;47:157-163.

397. Bulpitt CJ, Fletcher AE. Quality-of-life instruments in hypertension. PharmacoEconomics 1994;6:523-535.

398. Selby P. "Measuring the quality of life of patients with cancer". In: Quality of Life Assessment. Walker SR, Rosser RM, eds. Kluwer Academic Publishers, United Kingdom, 1993, pp. 235-267.

399. Maguire P, Selby P. Assessing the quality of life in cancer patients. Br J Cancer 1989;60:437.

400. de Haes JCJM, van Knippenberg FCE. The quality of life of cancer patients: a review of the literature. Social Sci Med

1985;20:809-17.

401. de Haes JCJM, van Knippenberg FCE, Neijt JP. Measuring psychological and physical distress in cancer patients: structure and application of the Rotterdam Symptom Checklist. Br J Cancer 1990;62:1034-38.

402. Bulpitt CJ, Fletcher AE. The measurement of quality of life in hypertensive patients: a practical approach. Br J Clin Pharmacol 1990;30:353-364.

403. Jones PW. "Measurement of health-related quality of life in asthma and chronic obstructive airways disease". In: *Quality of Life Assessment*. Walker SR, Rosser RM, eds. Kluwer Academic Publishers, United Kingdom, 1993, pp. 301-320.

404. Salek MS. "Measuring the quality of life of patients with skin disease". In: *Quality of Life Assessment*. Walker SR, Rosser RM, eds. Kluwer Academic Publishers, United Kingdom, 1993, pp. 355-370.

405. Deyo RA. "Measuring the quality of life of patients with rheumatoid arthritis". In: *Quality of Life Assessment*. Walker SR, Rosser RM, eds. Kluwer Academic Publishers, United Kingdom, 1993, pp. 269-287.

406. WHO. Tech Rep Ser 1969;425:5.

407. Karch FE, Lasagna L. Adverse drug reactions. JAMA 1975;234:1236.

408. Stephens MDB. The Detection of New Adverse Drug Reactions. Second Edition. Stockton Press, United States and Canada, 1988.

409. Skegg DCG, Doll R. The case for recording events in clinical trials. BMJ 1977;2:1523-1524.

410. Kramer MS, Leventhal JM, Hutchinson TA, Feinstein AR. An algorithm for the operational assessment of adverse drug reactions, I: Background, description for use and instruction. JAMA 1979;242:623-632.

411. Division of Drug Experience, Food and Drug Administration. Procedural Manual for Handling Drug Experience Reports. Glossary, Paper flow and Algorithms, 1980.

412. Bulpitt CJ. Screening for adverse drug reactions. Br J Hosp Med 1977;18:329-334.

413. Böttiger LE, Westerholm B. Drug-induced

blood dyscrasias in Sweden. BMJ 1973;3:339-343.

414. Multicentre international study. Reduction in mortality after myocardial infarction with long-term beta-adrenoceptor blockade. Supplementary Report. BMJ 1977;2:419-421.

415. Lewis JA. Pstmarketing surveillance: How many patients? Tips 1981, (April):93-94.

416. Vere DW. "Controlled trials to detect efficacy and toxicity: Training to meet tomorrow's needs". In: *Principles and practice of clinical trials*. Harris EL, Fitzgerald JE, eds. E & S Livingston, Edinburgh and London. 1970. pp 242-249.

417. Mann RD et al. "The significance of variations in the serum transaminases in the assessment of two new drugs". In: *Experimental studies and clinical experience - the assessment of risk* Proc Eur Soc for the Study of Drug Toxicity. Vol VI. Exerpta Medica Foundation, Amsterdam. 1965.

418. The coronary Drug Project Research Group. Clofibrate and niacin in coronary heart disease. JAMA 1975;231:360-381.

419. Knatterud GL, et al. Effects of hypoglycemic agents on vascular complications in patients with adult-onset diabetes. IV. A preliminary report on phenformin results. JAMA 1971;217:777-784.

420. Oliver MF. Serum cholesterol - the knave of hearts and the joker. Lancet 1981;2:1090-1095.

421. Dollery CT. "The risk identified from clinical trials". In: *Medicine and Risks/Benefit Decision*. Walker SR, Asscher AW, eds. MTP Press Limited, USA, 1987, pp. 57-65.

422. Bulpitt CJ, Fletcher AE. Cost-effectiveness of the treatment of hypertension. Clin Exp Hypertens 1993;15:1131-1146.

423. Johannesson M, Jönsson B. Cost-effectiveness analysis of hypertension treatment: a review of methodological issues. Health Policy 1991;19:55-78.

424. Drummond M. Economic Analysis Alongside Controlled Trials. Department of Health, 1994.

425. Maynard A. Logic in medicine: an economic perspective. BMJ 1987;295:1537-40.
426. Fletcher A. Pressure to treat and pressure to cost: a review of cost-effectiveness analysis. J Hypertens 1991;9:193-198.
427. Mehrez A, Gafni A. Quality-adjusted life-years, utility theory and healthy years equivalents. Medical Decision Making 1989;9:142-149.
428. Greenwood DT, Todd AH. "From laboratory to clinical use". In: *Clinical trials*. Johnson FN, Johnson S, eds. Blackwell Scientific Publications, Oxford. 1977. pp 13-35.
429. Simon TRM, Jones G. "Safety of medicines: The control of clinical trials". In: *Clinical trials*. Johnson FN, Johnson S, eds. Blackwell Scientific Publications, Oxford. 1977. pp 1-2.
430. Griffin JP, Long JR. New procedure affecting the conduct of clinical trials in the United Kingdom. BMJ 1981;283:477-478.
431. Baber NS. Volunteer studies: are current regulations adequate? the ethical dilemma. Pharmaceutical Medicine 1994;8:153-159.
432. Orme M, Harry J, Routledge P, Hobson S. Healthy volunteer studies in Great Britain: the results of a survey into 12 months' activity in this field. Br J Clin Pharmacol 1989;27:125-33.
433. Research on Healthy Volunteers. A Report of the Royal College of Physicians. The Royal College of Physicians, London. 1986;20:4.
434. Guidelines on the Practice of Ethics Committees in Medical Research involving Human Subjects (2nd edition). A Report of the Royal College of Physicians. The Royal College of Physicians, London, 1990.
435. Wells F. "Guidelines for medical experiments in non-patient human volunteers". In: *Medicines: Good Practice Guidelines*. ABPI, London, 1990:29-39.
436. Responsibility in Investigations on Human Participants and Material and on Personal Information. MRC Ethics Series, 1992.
437. Gillon R. No-fault compensation for victims of non-therapeutic research - should government continue to be exempt? J Med

Ethics 1992;18:59-60.

438. Watson N, Wyld PJ. The importance of general practitioner information in selection of volunteers for clinical trials. Br J Clin Pharmacol 1992;33:197-199.

439. Sutton A. Clinical pharmacology in development programmes: an essential ingrediant in good planning. Pharmaceutical Physician 1993;5:60-66.

440. Dollery CT. Clinical trials of new drugs. J Roy Coll Phys 1977;11:226-233.

441. Lumbroso A. "The introduction of new drugs". In: *Pharmaceuticals and health policy*. Blum R et al, eds. Croom Helm, London. 1981.

442. Crout JR. "Quoted in Lumbroso, A. The introduction of new drugs". In: *Pharmaceuticals and health policy*. Blum R, et al, eds. Croom Helm, London. 1981.

443. Silverman M, Lydecker M. "The promotion of prescription drugs and other puzzles. In: *Pharmaceuticals and health policy*. Blum R, et al, eds. Croom Helm, London. 1981. pp 78-92.

444. Lionel W, Herxheimer A. "Coherent policies on drugs: Formulation and implementation". In: *Pharmaceuticals and health policy*. Blum R, et al, eds. Croom Helm, London. 1981. pp240.

445. The Anturane Reinfarction Trial Research Group. Sulfinpyrazone in the prevention of sudden death after myocardial infarction. NEJM 1980;302:250-56.

446. Armitage P. Trials of antiplatelet drugs: some methodological considerations. Rev Epidemiol Sante Publique 1979;27:87-90.

447. Braunwald E. Treatment of the patient after myocardial infarction. The last decade and the next. NEJM 1980;302:290-92.

448. Kolata GB. FDA says no to Anturane. Science 1980;208:1130-32.

449. Mitchell JRA. Secondary prevention of myocardial infarction. The present state of the ART. BMJ 1980;2:1128-30.

450. Anturane Reinfarction Trial Policy Committee. The Anturane Reinfarction Trial: revaluation of outcome. NEJM 1982;306:1005-1008.

451. Anturane Reinfarction Italian Study Group. Sulphinpyrazone in post-myocardial infarction. Lancet 1982;i:237-42.
452. European Cooperative Study Group for Streptokinase Treatment in Acute Myocardial Infarction. Streptokinase in acute myocardial infarction. NEJM 1979;301:797-802.
453. Leading Article. Fibrinolytic therapy in myocardial infarction. BMJ 1979;2:1017-18.
454. Fletcher AP, Alkjaersig N, Smyrniotis FE, Sherry S. The treatment of patients suffering from early myocardial infarction with massive and prolonged streptokinase therapy. Trans Assoc Am Coll Physicians 1958;71:287-96.
455. Yusuf S, Wittes J, Friedman L. Overview of results of randomised clinical trials in heart disease. 1. Treatments following myocardial infarction. JAMA 1988;260:2088-93.
456. Elwood PC. "Aspirin, dipyridamole and secondary prevention". In: *The clinical impact of beta-adrenoceptor blockade*. Burley DM, Birdwood GFB, eds. Ciba Laboratories, Horsham, England, 1980. pp 25-26.
457. Elwood PC et al. A randomised controlled trial of acetyl salicylic acid in the secondary prevention of mortality from myocardial infarction. BMJ 1974;1:436-40.
458. Coronary Drug Project Research Group. Aspirin in coronary disease. J Chron Dis 1976;29:625-642.
459. Uberla K. In: *Acetylsalicylic acid in cerebral ischaemia and coronary heart disease. IV. Colfarit Symposium Berlin 1977*. Schattauer Verlag, Stuttgart and New York. 1978. P157.
460. Elwood PC, Sweetnam PM. Aspirin and secondary mortality after myocardial infarction. Lancet 1979;2:1313-15.
461. The Persantine-Aspirin Reinfarction Study Research Group. Persantine and aspirin in coronary heart disease. Circulation 1980;62:449-61.
462. Aspirin Myocardial Infarction Study Group. A randomised, controlled trial of aspirin in

persons recovered from myocardial infarction. JAMA 1980;243:661-69.

463. Antiplatelet Trialists' Collaboration. Secondary prevention of vascular disease by prolonged antiplatelet treatment. BMJ 1988;296:320-31.

464. Doubts about dipyridamole as an antithrombotic drug. Drug and Therapeutics Bulletin 1984;22:26-28.

465. Gent AE, Brook CGD, Foley TH, Miller TN. Dipyridamole: a controlled trial of its effect in acute myocardial infarction. BMJ 1968;4:366-8.

466. Klimt CR. Knatterud GL, Stamler J, Meier P. Persantine-aspirin reinfarction study. Part II. Secondary coronary prevention with persantine and aspirin. J Am Coll Cardiol 1986;7:251-69.

467. Fitzgerald GA. Dipyridamole: NEJM 1987;316:107-117.

468. Sixty Plus Reinfarction Study Research Group. A double-blind trial to assess long-term oral anticoagulant therapy in elderly patients after myocardial infarction. Lancet 1980;2:989-993.

469. Chalmers TC, Matta RJ, Smith H Jr, Kunzler AM. Evidence favouring the use of anticoagulants in the hospital phase of acute myocardial infarction. NEJM 1977;297:1091-6.

470. Peto R. Clinical trial methodology. Biomed Pharmacother 1978;28(special issue):24-36.

471. Mitchell JRA. Anticoagulants in coronary heart disease; Retrospect and prospect. Lancet 1981;i:257-62.

472. Smith P, Arnesen H, Holme I. The effect of warfarin on mortality and reinfarction after myocardial infarction. NEJM 1990;323:147-52.

473. Vedin JA. Analysis presented at the VIIIth European Congress of Cardiology. Paris, 1980.

474. Reynolds JL, Whitlock RM. Effects of beta-adrenergic receptor blocker in myocardial infarction for one year from onset. Br Heart J 1972;34:252-59.

475. Ahlmark G, Saetre H. Long-term treatment with ß-blocker after myocardial infarction.

424

Eur J Clin Pharmacol 1976;10:77-83.
476. Wilhelmsson C, et al. Reduction of sudden deaths after myocardial infarction by treatment with alprenolol. Preliminary results. Lancet 1974;2:1157-1160.
477. Anderson MP et al. Effect of alprenolol on mortality among patients with definite or suspected acute myocardial infarction. Preliminary results. Lancet 1979;1:865-68.
478. Hampton JR. "Evidence on use of beta-blockers". In: *Current themes in cardiology*. Birdwood GFB, Russel JG, eds. Geigy Pharmaceuticals, Horsham, England. 1981. pp.86-89.
479. Beta blocker heart attack study group. The beta blocker heart attack trial. JAMA 1981;246:2073-74.
480. Held P, Yusuf S, Furberg CD. Calcium channel blockers in acute myocardial infarction and unstable angina: an overview. BMJ 1989;299:1187-92.
481. Furberg CD, Psaty BM, Meyer JV. Nifedipine. Dose-related increase in mortality in patients with coronary heart disease. Circulation 1995;92:1326-1331.
482. Schor S. The university group diabetes program. A statistician looks at the mortality results. JAMA 1971;217:1671-75.
483. Feinstein AR. An analytic appraisal of the university group diabetes program (UGDP) study. Clin Pharmacol Ther 1971;12:167-91.
484. Data has holes, statistical men report to USV. Drug Trade News, 23 August 1971.
485. Cornfield J. The University Group Diabetes Program. A further statistical analysis of the mortality findings. JAMA 1971;217:1676-87.
486. Committee for the assessment of biometric aspects of controlled trials of hypoglycemic agents. JAMA 1975;231:585-608.
487. Wihelmsen L, Berglund G, Elmfeldt D et al. Beta-blockers versus diuretics in hypertensive men: Main results from the HAPPHY trial. J Hypertens 1987;6:561-572.
488. Wikstrand J, Warnold I, Olsson G, et al. Primary prevention with metoprolol in patients with hypertension: Mortality

results from the MAPHY Study. JAMA 1988;269:1976-1982.

489. Wihelmsen L. Primary prevention with metoprolol in patients with hypertension (letter). JAMA 1988;260:1713.

490. Staessen J, Fagard R, Amery A. Primary prevention with metoprolol in patients with hypertension (letter). JAMA 1988;260:1713-1714.

491. Wikstrand J, Warnold I, Olsson G, Tuomilehto J, Elmfeldt D, Berglund G (Advisory Committee for the MAPHY Study). Primary prevention with metoprolol in patients with hypertension (letter). JAMA 1988;260:1715-16.

492. Gifford RW. Primary prevention with metoprolol in patients with hypertension (letter). JAMA 1988;260:1714-1715.

493. Johannesson M, Wikstrand J, Jönsson B, Berglund G, Tuomilehto J. Cost-effectiveness of antihypertensive treatment. Metoprolol versus thiazide diuretics. PharmacoEconomics 1993;3:36-44.

494. EC/IC Bypass Study Group. Failure of extracrandial-intracranial bypass to reduce the risk of ischemic stroke: results of an international randomised study. NEJM 1985;313:1191-200.

495. Dudley HAF. Extracranial-intracranial bypass, one; clinical trials, nil. BMJ;294:1501-2.

496. Sundt T. Was the international randomised trial of extracranial-intracranial arterial bypass representative of the population at risk? NEJM 1987;316:814-6.

497. Relman A. The extracranial-intracranial arterial bypass study - what have we learnt? NEJM 1987;316:820-4.

498. Douglas-Jones AP, Cruickshank JM. Once daily dosing with atenolol in patients with mild or moderate hypertension. BMJ 1976;1:990-91.

499. Bulpitt CJ, Rose GA. "Design and conduct of clinical trials". In: *The hypertensive patient*. Marshall AJ, Barrett DW, eds. Pitman Medical, Tunbridge Wells. 1980. pp494-505.

500. Harris WH, et al. Aspirin prophylaxis of

426

venous thromboembolus after total hip replacement. NEJM 1977;297:1246-1249.

501. Cranberg L. Do retrospective controls make clinical trials 'inherently fallacious'? BMJ 1979;2:1265-6.

502. Leading article. Randomised controlled trials? BMJ 1979;4:1244-5.

503. Rogers SC, Clay RM. A statistical review of controlled trials of imipramine and placebo in the treatment of depressive illness. Br J Psychiat 1975;127:599-603.

504. Maxwell C. Clinical trials, reviews, and the journal of negative results. Br J Clin Pharmacol 1981;1:15-18.

505. Mittra B. Potassium, glucose and insulin in treatment of myocardial infarction. Lancet 1965;2:607-9.

506. Hargreaves MA, Maxwell C. The speed of action of desipramine: a controlled trial. Intl J Neuropsychiat 1961;3:140-1.

507. Fincke M. Arzneimittelprüfund: Strafbare Versuchsmethoden. Heidelberg/Karlsruhe, 1977.

508. Leading article. Controlled tials: planned deception? Lancet 1979;1:534-5.

509. Goldzieher JW, et al. A double-blind cross-over investigation of the side effects attributed to oral contraceptives. Fertil Steril 1971;22:609-623.

510. Pryce IG. Clinical research upon mentally ill subjects who cannot give informed consent. Br J Psychiat 1978;133:366-69.

511. Black D. The paradox of medical care. J Roy Coll Phys Lond 1979;13:57-65.

512. Cochrane AL. Attitudes to controlled trials. Paper presented at the 9th International Scientific Meeting of the IEA. Edinburgh, August 1981.

INDEX